D0578010

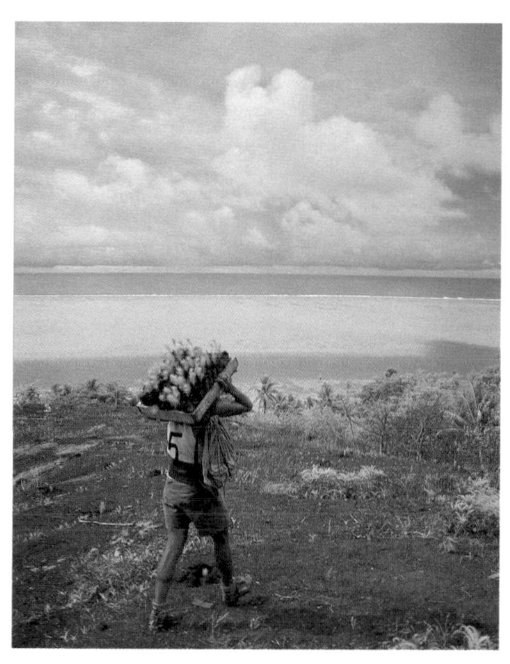

GLOBAL WARMING

AMERICAN MUSEUM OF NATURAL HISTORY

ENVIRONMENTAL DEFENSE FUND

GLOBAL WARMING

UNDERSTANDING THE FORECAST

ANDREW REVKIN

ABBEVILLE PRESS PUBLISHERS ■ NEW YORK LONDON PARIS

Editor: Susan Costello
Designer: Renée Khatami
Production Supervisor: Hope Koturo
Production Editor: Cristine Mesch
Picture Researchers: Linda Lilienfeld, Barbara Becker, Christa Kelly
Illustrator: Paul Singer

Jacket front: View of a jet stream over the Nile Valley and the Red Sea (NASA).
Jacket back: Polar bear in an arctic landscape (Hans Reinhard/Okapia/Photo Researchers, Inc.).
Page 1: A worker carrying tree seedlings to be planted on the island of Raiatea, French Polynesia (Andrew Revkin).
Pages 2–3: A smoldering forest fire in the Brazilian Amazon (Loren McIntyre).
Page 8: Crabeater seals rest on sea ice in Antarctica (Robert W. Hernandez/Photo Researchers, Inc.).

Copyright © 1992 American Museum of Natural History and Environmental Defense Fund. All rights reserved under international copyright conventions. No part of this book may be reproduced or utilized, in any form or by any means, electronic or mechanical, including photocopying, recording, or by any information storage and retrieval system, without permission in writing from the publisher.
Inquiries should be directed to Abbeville Press, 488 Madison Avenue, New York, NY 10022. Printed and bound in Italy. First edition.

Library of Congress Cataloging-in-Publication Data
Revkin, Andrew.
Global warming : understanding the forecast / Andrew Revkin.
p. cm.
At head of title: American Museum of Natural History, Environmental Defense Fund.
Includes bibliographical references (p.) and index.
ISBN 1-55859-310-1
1. Global warming. 2. Climatic changes. I. American Museum of Natural History. II. Environmental Defense Fund. III. Title.
QC981.8.G56R48 1992
363.73'87—dc20 91-35720

Printed on recycled paper.

In cosponsoring this book and the associated traveling exhibition on global climate change, the American Museum of Natural History and the Environmental Defense Fund have taken a bold step in focusing public attention on an issue of increasing concern. On behalf of the exhibition's Steering Committee, I wish to thank those who have given such generous support and encouragement to our efforts, including the major contributors listed below. We hope the exhibition and this book will increase public understanding of the difficult environmental challenges that lie ahead.

Frank E. Taplin, Jr.
Chairman, Steering Committee

National Science Foundation
John D. and Catherine T. MacArthur Foundation
The Annenberg Foundation
Atlantic Foundation
The William Bingham Foundation
Mr. and Mrs. Leonard N. Block
Janet and Richard D. Colburn
Geraldine R. Dodge Foundation, Inc.
Charles H. Dyson
The Armand G. Erpf Fund, Inc.
The Horace W. Goldsmith Foundation
Mary A. and Thomas F. Grasselli Foundation
The Greenville Foundation
Heineman Foundation for Research, Educational, Charitable, and Scientific Purposes, Inc.
Teresa Heinz and sons, in memory of United States Senator John Heinz
W. Alton Jones Foundation, Inc.
George and Elinor Montgomery
Richard Pratt
Mr. and Mrs. David Rockefeller
Mr. and Mrs. William H. Scheide
Mr. and Mrs. Frank E. Taplin, Jr.
Turner Broadcasting System, Inc.

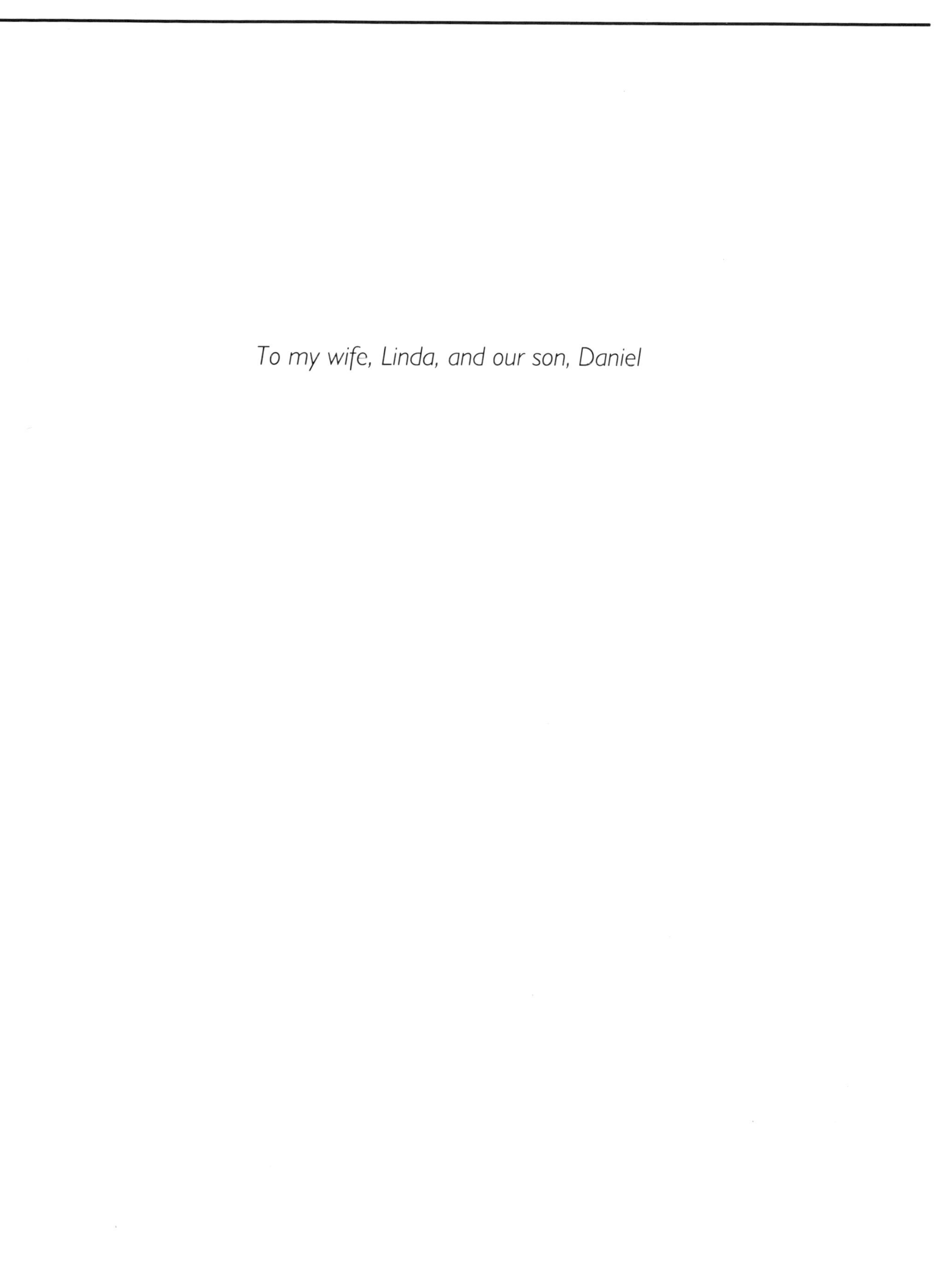

To my wife, Linda, and our son, Daniel

PREFACE by Fred Krupp, Executive Director, Environmental Defense Fund, and George D. Langdon, Jr., President, American Museum of Natural History 10

AN ICE ROAD ACROSS THE BAY 12

A SCENE OF CHANGES 30

THE GLOBAL GREENHOUSE 56

CONTENTS

THE HAND OF MAN 74

THE CLOUDY CRYSTAL BALL 100

BUSINESS AS USUAL 116

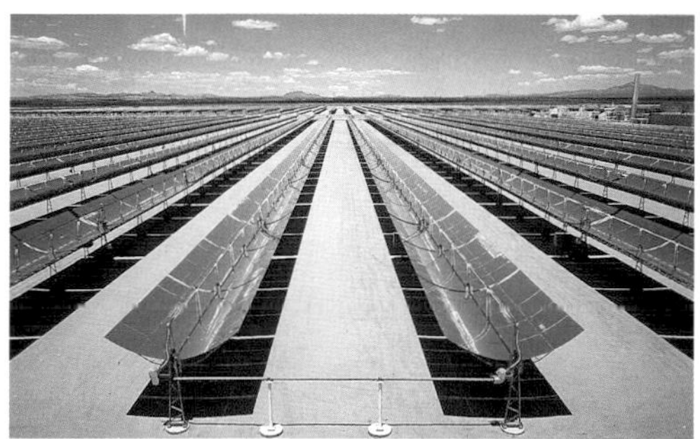

CHOOSING OUR FATE 140

Suggested Reading 167

A GREENHOUSE DIET: 20 Things You Can Do to Reduce the Risk of Global Warming 169

Organizations and Resources 173

Index 177

Picture Credits 180

I am grateful to many people for their help in seeing this book into print. First is my wife, Linda Lieff, who, as always, kept me from faltering whenever I felt I was losing ground. Our son, Daniel, was born just a couple of months before this book was conceived; as a member of the generation that will inherit the consequences of our actions, he helped add urgency to my task.

At the American Museum of Natural History, I thank Eva Zelig for proposing that I write the book—and for her determination and optimism. Scarlett Lovell provided valuable suggestions. Thomas Kelly, the museum's publisher, kept the project moving ahead full speed.

The manuscript was studiously critiqued by Malcolm McKenna and George Harlow, at the American Museum of Natural History, and John Firor, at the National Center for Atmospheric Research. Wallace Broecker, at Columbia University's Lamont-Doherty Geological Observatory, Thomas Webb, at Brown University, and Stanley

Awramik, at the University of California at Santa Barbara, added helpful comments.

Special thanks go to several people at the Environmental Defense Fund, whose comments and ideas greatly improved this book. The careful scrutiny of Michael Oppenheimer and Stephanie Pfirman was indispensable. Sarah Clark developed the practical actions found in "A Greenhouse Diet." Joel Plagenz, Roger Pasquier, Daniel Dudek, and Alice Le Blanc all contributed useful thoughts. Brian O'Neill was tireless in his efforts to ensure that the text and graphics were correct and effective. Any errors that survived this gauntlet and appear in the book are my responsibility alone.

Linda Lilienfeld, the photo researcher, assisted by Barbara Becker and Christa Kelly, did a superb job of rounding up images and injected the book with a dose of her unique effervescence. At Abbeville Press, Renée Khatami's creative eye produced a compelling design. Cristine Mesch, the production editor, and Hope Koturo, the production supervisor, also deserve special recognition for their caring and skillful contribution.

Finally, Susan Costello, Abbeville's managing editor, was passionate about this book, and her passion shows through in the final product.

Andrew Revkin

Can humans change the climate? Could such everyday actions as turning on the lights and stepping on the gas have the unintended effect of warming our planet? What difference would a few degrees make, anyway?

It's easier to pose such questions about global warming than to find complete answers, because the forces that determine climate are so complex. Understanding the forecast means, above all, understanding the *process* by which scientists investigate the past and present in order to anticipate the future.

Since its founding in 1869, the American Museum of Natural History has been a leader in education and research in the fields of natural science and anthropology. From fossil records that date back millions of years, the Museum's scientists have studied how life on Earth has evolved amid occasional dramatic changes in global climate. Now the Museum has joined with the Environmental Defense Fund—a member-supported group of scientists, economists, and attorneys who develop solutions to environmental problems—in an unprecedented collaboration on the issue of global warming. This jointly published book, and the traveling exhibition that it complements, share a common mission: to promote understanding of Earth's climate and to show how humans have acquired the power to change it.

Andrew Revkin weaves a story in which scenes of climate—past, present, and future—appear vividly before us. We are taken to hotter times when crocodiles roamed the Arctic and colder times when the Manhattan site of the American Museum was buried under a thousand feet of ice. In a compelling scenario, we are then shown one possible future that may result if we continue in our present course of "Business As Usual."

Current human activities—such as the widespread burning of fossil fuels to run power plants and vehicles—are releasing carbon dioxide and other "greenhouse" gases into the atmosphere (gases so named because they trap heat much as the glass in a greenhouse does). If present trends continue and these gases effectively double in concentration during the coming century, a committee of the National Academy of Sciences predicts that global average temperatures would rise between 3° and 8°F (between 1.5° and 4.5°C), leaving the Earth warmer than at any time in human history. Revkin points out that a rise of only 9°F in global average temperature terminated the last ice age.

Scientists are still trying to sort out the possible consequences of such future warming. Were it to occur suddenly, it could disrupt human societies that are already stressed by limited resources and rapid population growth. Living things of all kinds could be imperiled in their effort to keep pace with shifting climate zones.

Yet the risk of global warming could be reduced by our actions today, particularly by how thoroughly we employ energy efficiency and renewable energy sources—options that are worth considering for other reasons as well.

These issues must be addressed not only at the highest levels of international diplomacy but also in our day-to-day lives. With the average American responsible for 40,000 pounds of annual carbon dioxide emissions, the Greenhouse Diet at the back of this book offers each of us a constructive opportunity for change—through actions as simple as planting a shade tree or flipping the "Energy Saver" switch on a home appliance.

Our power to affect the Earth's climate may have outstripped our ability to understand and predict the consequences. It's time to catch up and prepare to adjust our actions accordingly. There is no other responsible course.

By Fred Krupp, Executive Director, Environmental Defense Fund, and George D. Langdon, Jr., President, American Museum of Natural History

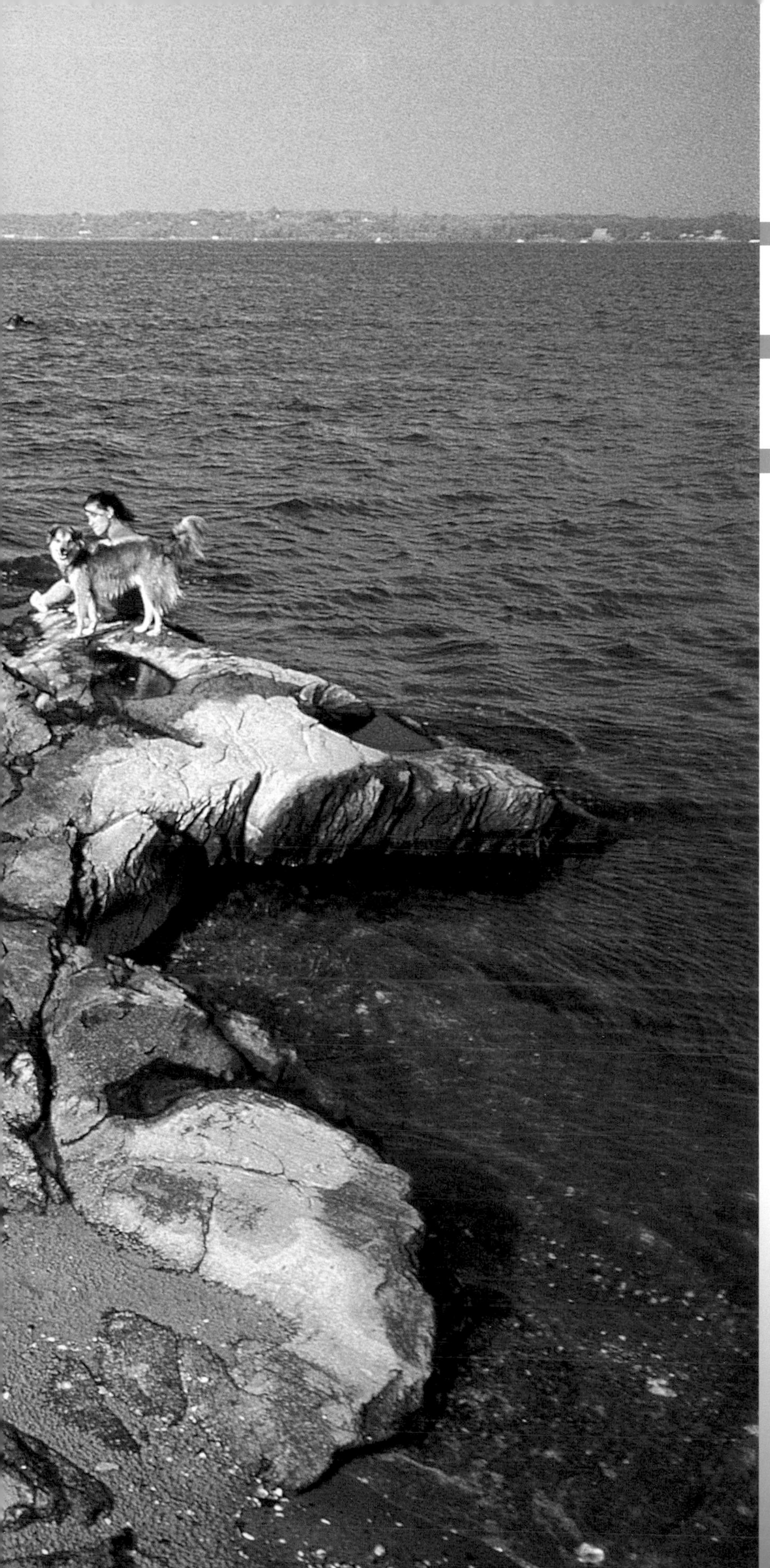

AN ICE ROAD ACROSS THE BAY

"It is hard to feel affection for something as totally impersonal as the atmosphere, and yet there it is, as much a part and product of life as wine or bread."

—Lewis Thomas,
physician and author (b. 1913),
The Lives of a Cell

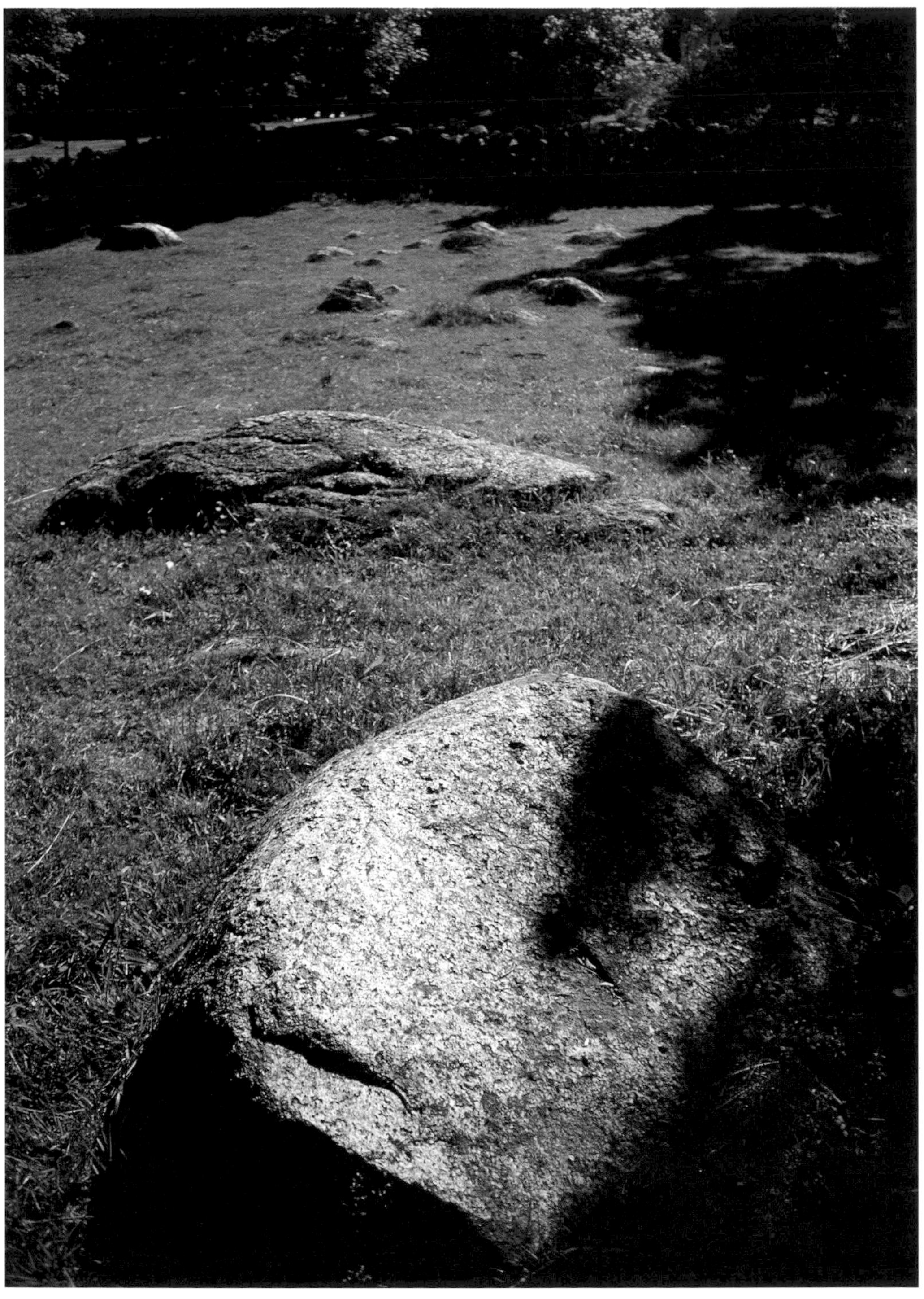

AN ICE ROAD ACROSS THE BAY

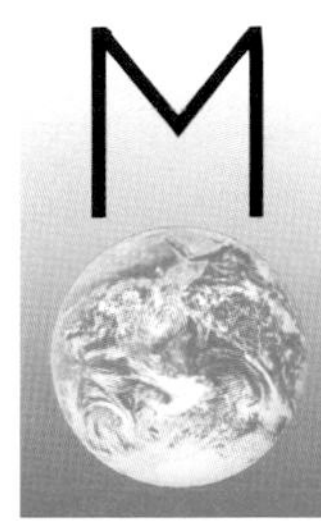

My parents live on Sally Rock Point, a little wooded spit that juts into a branch of Narragansett Bay, the waterway that splits Rhode Island up the middle. I often walk down from their cottage to a flat shield of shale that meets the waves at the point's end. It's a quiet spot to sit and think. Gulls and an occasional red-tailed hawk soar overhead, and hermit crabs scuttle across the white field of barnacles that paints the rock below the high-tide line. Fishermen buzz by in their skiffs, but not much else goes on. The nearest town is on the opposite shore, across more than a mile of water.

Before the turn of the century, it was common each winter for coal wagons to take a short cut from the far side of the bay to the homes on Sally Rock Point. The wagons were driven across the thick ice that formed over the entire expanse of calm salt water. More recently, in every winter I have known, there has been no ice road across the bay. Ice still forms along the shores, and sometimes a thin sheet forms briefly over the whole bay, but it is never so thick that you could walk on it, let alone drive a truck across it.

The warmth that has prevented the bay from freezing recently may simply be a fluke of New England weather, the changeability of which is legend. Then again, the milder winters may be a sign that it is not just the weather that is changing this time, but the *climate*—the general pattern of temperature and moisture for the region. And this regional warming may be a small part of a global trend.

Climate change is nothing new. Evidence of this can be found in the boulders strewn around the pastures in Exeter, a few miles inland from Sally Rock Point. These refrigerator-size chunks of granite were deposited by a half-mile-thick sheet of ice that scraped south across New England 20,000 years ago. At the time—because of slow, regular variations in Earth's orbit and other factors—ice covered vast regions of North America, Europe, and Asia. Standing on Sally Rock Point today, it is difficult to envision the time when thousands of feet of ice pressed down on the land. It is also remarkable

Climate is always changing—a point illustrated by the boulders studding a pasture in Exeter, Rhode Island (opposite). They were left by a towering glacier that scraped south over New England just 20,000 years ago. The expected warming from the greenhouse effect may be nearly as dramatic as the temperature change that ended that ice age.

Rhode Island's Sally Rock Point (preceding pages) is just one place that could suffer the changes in climate and sea level that may accompany global warming. Rising seas may inundate the scenic spot. Warmer temperatures may disrupt life on land and in the water.

to think that the average temperature on the planet has risen just 9°F (5°C) since that time—that just 9°F can mean the difference between a half-mile of ice and a wind-dappled bay with a forested shore.

The Warming Atmosphere

The climate change that may be occurring now is disturbingly different from the slow, steady cycle of ice ages and warmings that has sculpted the face of Earth for two million years. The disappearance of sea ice from this arm of Narragansett Bay may be one result of the warming of Earth's atmosphere by an increasing greenhouse effect. The greenhouse effect is the tendency of certain gases in the atmosphere to trap heat in much the same way that the glass panes of a greenhouse roof help make it possible to grow tomatoes in winter.

Earth's atmosphere has always acted like a greenhouse. Water vapor and a tiny trace of carbon dioxide—just a few hundredths of a percent—allow sunlight in but prevent the sun-warmed planet from radiating all that energy back into space. Indeed, without this insulating blanket, Earth would more closely resemble its frozen, barren cousin Mars, whose thin atmosphere retains much less heat. Now, for the first time, scientists are concerned that the composition of Earth's atmosphere is being rapidly altered by human activity.

It may be just a fluke of New England weather, or one bit of evidence that global warming is under way: Rhode Island's Narragansett Bay once froze so thick that carts could be wheeled across the ice. Lately, however, it has remained largely ice-free. In a photograph taken in April 1893, workers harvest oysters through holes cut in the sea ice.

In a way, it is not surprising that a species as prolific and industrious as *Homo sapiens* should have an impact on the dynamics of the entire globe. Since the last time the ice sheets retreated toward the poles, some 15,000 years ago, the number of humans on the planet has risen from less than 5 million to more than 5 billion. Even if human populations only modified the landscape in the simplest ways, say, by chopping down forests, the effect on the planet would be significant. But the human impact has been amplified to an extraordinary degree not only by our numbers but also by our ability to fashion tools and technologies that increase our power to change the world. Here is a species that began by taming fire and has since learned to replicate the fusion energy of the sun in a hydrogen bomb.

Along the way, humans discovered the vast stores of energy locked up in subterranean pockets of oil, coal, and natural gas—the fuels that stoked the boilers of the Industrial Revolution and still power our productive but profligate life-style today. Just since World War II, the economic output of industrialized countries has increased forty times over. But there has been a hidden cost. All of that combustion—in power plants and automobiles and factories—has transformed hundreds of billions of tons of ancient, buried carbon into a great burst of carbon dioxide gas that has significantly changed the atmosphere. The incineration of tropical forests, by releasing more carbon dioxide, has added greatly to the problem.

A Burning Issue

Today, for every one of the 5.3 billion people on Earth, nearly four tons of carbon dioxide are spewed into the air annually. In the energy-addicted United States, the rate is almost five times as high. Americans consume 22 percent of the world's oil, even though they make up just 5 percent of the world's population. We live at a time when one person commutes to work in an automobile that typically expends the energy consumed by 140 horses: In a year, a typical commuter's car burns so much gasoline that it releases more than three times its own weight in carbon dioxide into the atmosphere.

As a result, in just the past hundred years, the atmospheric concentration of this heat-trapping gas has risen 22 percent. By the latter half of the next century, it is likely that the amount of carbon dioxide in the atmosphere will have doubled from preindustrial levels.

Moreover, other gases generated by human agriculture and industry also trap heat—gases such as methane, nitrous oxide, and chlorofluorocarbons, or CFCs. (Those same CFCs, used as refrigerants, propellants in some spray cans overseas, and in some foam packaging, also attack the protective shield of ozone in the upper reaches of the atmosphere.) Overall, the warming effect of these other greenhouse gases is expected eventually to equal, if not exceed, that of carbon dioxide.

Thus an era has begun in which humans are no longer simply polluting a particular lake, or cutting down a certain forest, but changing the composition and dynamics of one of the essential components of the planet. Because the atmosphere is intimately linked with Earth's other components—the oceans, the soil, the sheets of ice at the poles, the flow of energy from the sun, and the web of life—humans have, in an instant of geological time, taken hold of the reins that will guide this rare blue sphere into the future.

Many atmospheric scientists predict that some time in the present decade, we shall detect the long-anticipated climatic "signal"—clear evidence that all of these emissions from human activity have turned up the global thermostat. In northern latitudes, leaves will still fall in October and snow will still fall in February. But, by many accounts, Washington, D.C., and Dallas will be more likely to have a particularly steamy summer; the Grain Belt will have increased chances of a drought; and the odds of the Florida Keys being ravaged by a severe hurricane—and the likelihood of its winds being more destructive—will have risen. Already in the decade of the 1980s, Earth experienced the six hottest years on record. The first year of the 1990s was hotter still.

Looking Ahead

Computer models that simulate the workings of the atmosphere project that the doubled level of carbon dioxide will raise the world's average temperature anywhere from 3° to 8°F (1.5° to 4.5°C). In other words, it is possible that our climate will be jogged by a change nearly as dramatic as the one that ended the last ice age. This new change, though, will occur in just one century, not fifty centuries. In that instant of geological time, the planet may become warmer than it has been in several million years.

Many of the predicted impacts of such a change bode ill for humanity and wildlife: The warming oceans may expand and glaciers melt, with the resulting rise in sea level inundating coastal communities and creating millions of eco-refugees; changing climate patterns will likely disrupt agriculture; ecosystems that cannot shift fast enough to keep up may be exterminated; the frozen tundra of the far north may thaw, releasing massive amounts of methane and thus exacerbating the greenhouse effect. In 1987, the list of predictions filled a heavy red book—a book as large as the Manhattan Yellow Pages

An aerial view (opposite) shows a fire roaring through Brazil's Amazon rain forest during the annual burning season. The accelerating destruction of rain forests around the world contributes to the greenhouse effect by releasing large amounts of carbon dioxide as vegetation burns.

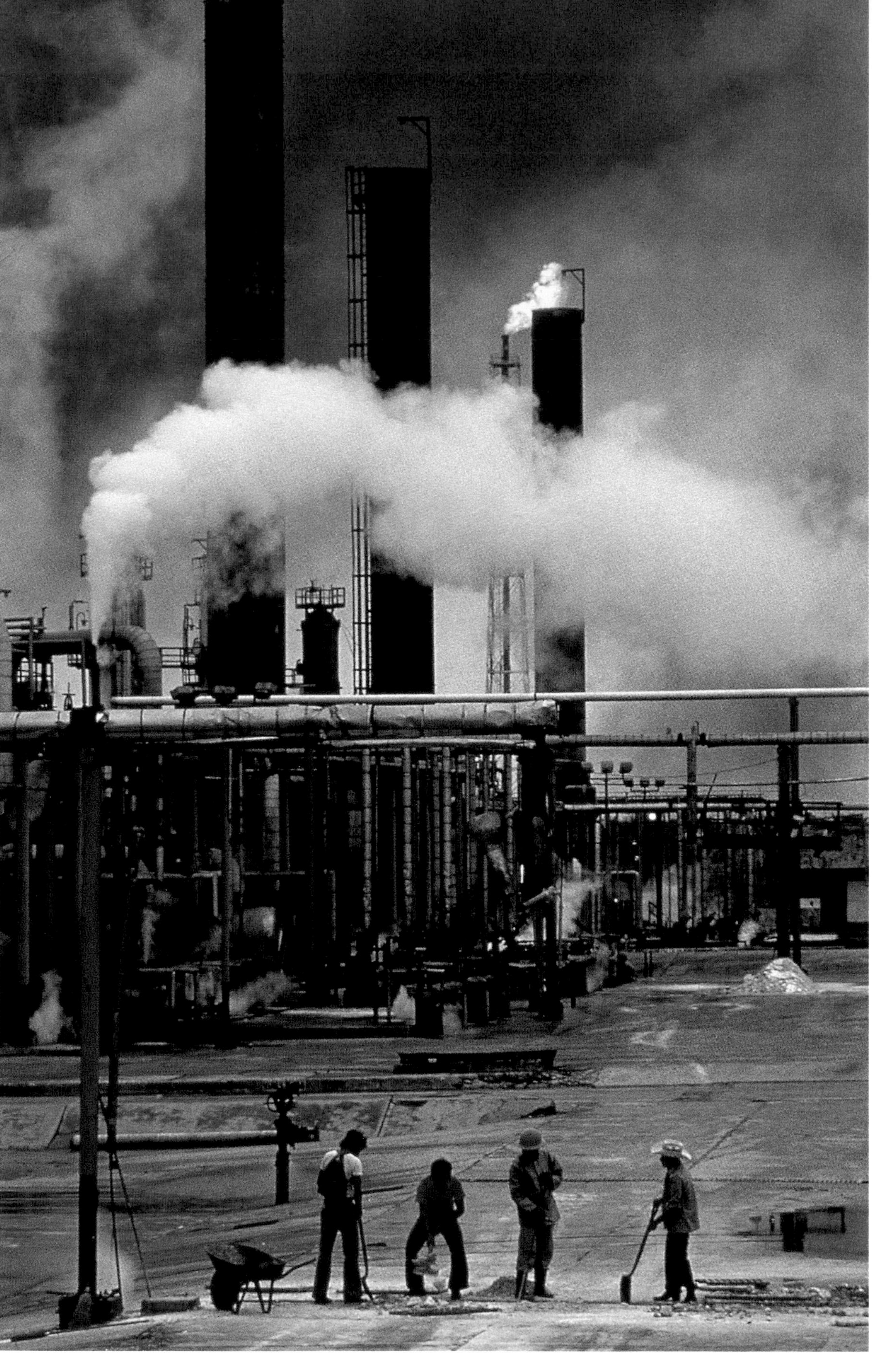

A refinery in Chiapas, Mexico (opposite), turns crude oil into gasoline and other products that fuel modern industrial society. Much of the heat-trapping carbon dioxide being added to the air by humans comes from the burning of petroleum products.

A factory in Romania (right) spews enormous amounts of carbon dioxide and noxious compounds. For decades, industries in Eastern Europe have operated without environmental regulations.

Pollution stains the air over Calais, France (left). Since World War II, the industrial production of developed nations such as France and the United States has grown forty-fold.

Children waiting to enter their school in a Czechoslovakian village wear masks to protect them from severe air pollution. Some of the greatest opportunities for reducing carbon dioxide emissions and toxic pollution are in Eastern Europe.

—called *Preparing for Climate Change*. In the 1990s, the heavy tomes are coming thick and fast, focusing on everything from the spread of insect-borne diseases to the deterioration of coral reefs.

Fortunately, the same intelligence that has allowed humans to dominate and scar the planet in such a short time also endows us with the ability to anticipate the consequences of our actions. Hundreds of scientists worldwide have made clear the import of our current course. But foresight, in itself, is insufficient. Scientists have suggested dozens of prudent actions that can be taken now to limit the impending disruption to both civilization and the biosphere—actions ranging from using a more efficient light bulb to planting a forest. If evidence for global warming continues to mount, more dramatic measures may spell an early end to the age of oil and coal, when progress came so cheaply—mined or pumped from a hole in the ground. For industrialized nations, such measures would include modifying the yardstick by which progress is traditionally calculated—taking into account, for the first time, the environmental cost of growth. For the developing nations, global warming may bring an unusual opportunity—to leapfrog past the mistakes of the industrialized world and create economies that are attuned to the needs of both people *and* the planet.

Unfortunately, the changes in the atmosphere, although racing along at a pace unprecedented in recent planetary history, are imperceptible to human eyes. Any signal of global warming is still largely hidden in the statistical "noise" produced by normal fluctuations in weather. The lack of hard evidence has caused some politicians and pundits to resist cuts in emissions of greenhouse gases, for fear that economic growth will be imperiled. We can afford to wait a while, they say. But among climate experts and earth scientists, a strong consensus is emerging that the evidence for global warming is sufficient to justify action now. As Michael McElroy, chairman of Harvard University's Department of Earth and Planetary Sciences, has said, "If we choose to take on this challenge, it appears that we can slow the rate of change substantially, giving us time to develop mechanisms so that the cost to society and the damage to ecosystems can be minimized. We could alternatively close our eyes, hope for the best, and pay the cost when the bill comes due."

Another valuable perspective is provided by José Lutzenberger, a noted environmentalist and Brazil's first Secretary of the Environment. His appointment in 1990 was a hopeful development for Brazil, a nation that had incinerated an area of Amazon rain forest twice the size of California in just ten years. One evening, Lutzenberger and I sat in a town deep in the rain forest there, in a region where it is rare not to smell wood smoke in the wind—smoke from thousands of fires set by people clearing the jungle to make cattle pasture. "A complicated system can take a lot of abuse, but you get to a point where suddenly things fall apart," Lutzenberger said. "It's like pushing a long ruler toward the edge of a table. Nothing happens, nothing happens, nothing happens. Then, suddenly, the ruler falls to the floor." That may well be true for climate: By the time the ruler clatters to the floor, it may be too late.

"Managing" the Atmosphere

When I was a college student in London some fourteen years ago, I stopped by one day at a little booksellers' fair that convened every lunch hour in the financial district. Among the crumbling leather-bound remains of someone's library, piled high on a wooden cart, I found a volume called *The Physical Geography of the Sea*, by Matthew Fontaine Maury. It was a sea captain's guide to the basics of oceanography and meteorology, published in 1859 by Sampson Low, Son, and Co. The book sat largely unread until this year, when I opened it and found a chapter entitled "The Atmosphere."

Nowhere else have I seen a passage that so effectively describes the workings of the "spherical shell which surrounds our planet," as the author

The biggest increase in average temperatures is expected in the Arctic, home to such species as the northern seal (opposite), seen here on Alaska's Afognak Island.

In a warmer world, animals of ecosystems such as the African savanna may be forced to shift their ranges to keep up with rapidly changing patterns of rainfall. As a result, many vulnerable species may face extinction. Above, a herd of African elephants searches for water in the dry season. At right, impalas flee from a lion.

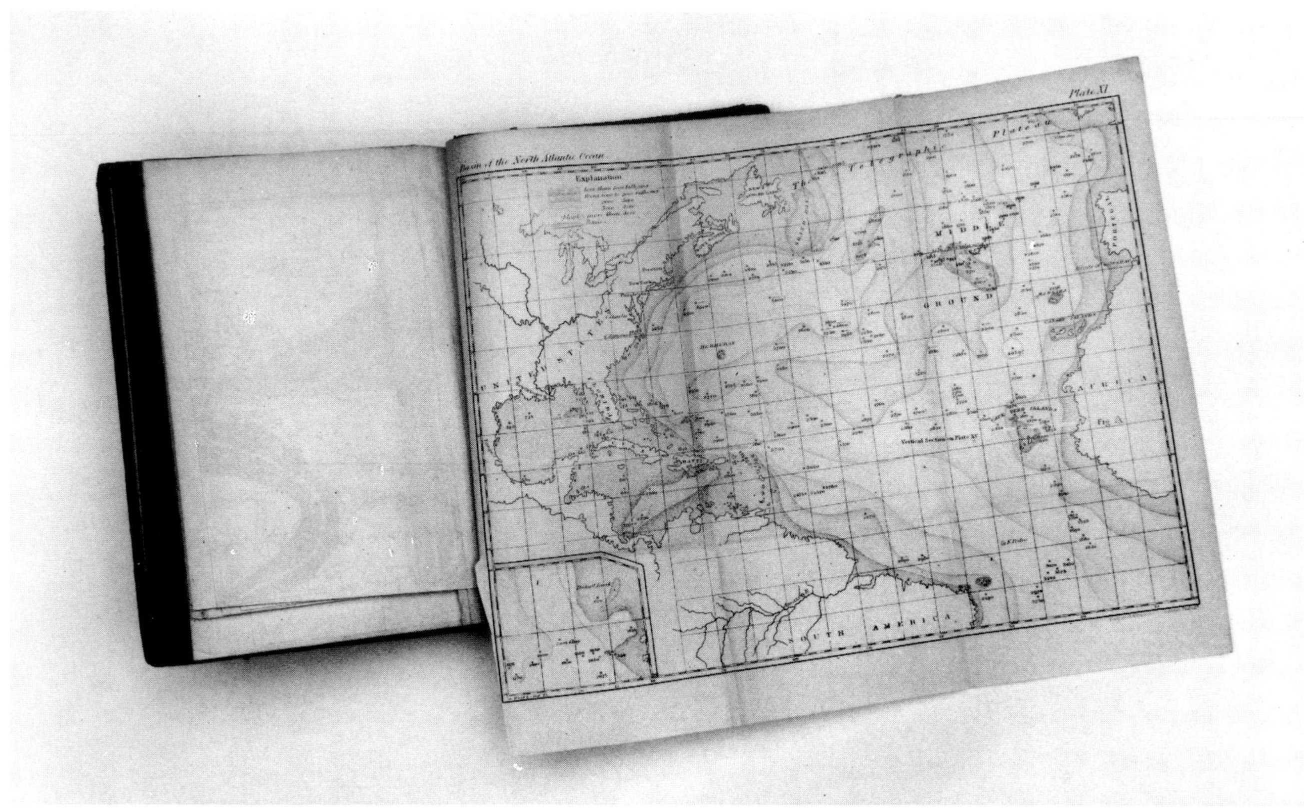

M. F. Maury's *The Physical Geography of the Sea*, published in 1859, contains passages that powerfully describe the workings of the atmosphere—and the importance of treating it well. This is a fold-out chart of the North Atlantic.

puts it. The book speaks powerfully of the importance of treating the atmosphere with respect. The atmosphere

warms and cools by turns the earth and the living creatures that inhabit it. It draws up vapours from the sea and land, retains them dissolved in itself, or suspended in cisterns of clouds, and throws them down again as rain or dew when they are required. . . . It affords the gas which vivifies and warms our frames, and receives into itself that which has been polluted by use, and is thrown off as noxious. . . .

It is only the girdling encircling air, that flows above and around all, that makes the whole world kin. The carbonic acid [carbon dioxide] with which to-day our breathing fills the air, to-morrow seeks its way round the world. The date-trees that grow round the falls of the Nile will drink it in by their leaves . . . and the palms and bananas of Japan will change it into flowers. The oxygen we are breathing was distilled for us . . . by the magnolias of the Susquehanna, and the great trees that skirt the Orinoco and the Amazon. . . . The rain we see descending was thawed for us out of the icebergs which have watched the polar star for ages, and the lotus lilies have soaked up from the Nile, and exhaled as vapour, snows that rested on the summits of the Alps.

Hence, to the right-minded mariner, and to him who studies the physical relations of earth, sea, and air, the atmosphere is something more than a shoreless ocean, at the bottom of which he creeps along. . . . It is an inexhaustible magazine, marvellously adapted for many benign and beneficent purposes.

This photograph, taken by Apollo astronauts on a 1972 voyage to the moon, shows the unity of the atmosphere, with great swirling features that span thousands of miles. Antarctica is at the bottom, with much of Africa visible.

The sky over Rhode Island's Sally Rock Point glows at sunset. The atmosphere appears unchanging, but it has been significantly altered by human actions.

Upon the proper working of this machine depends the well being of every plant and animal that inhabits the earth; therefore the management of it, its movements, and the performance of its offices, cannot be left to chance.

Now we have arrived at a time when, voluntarily or involuntarily, humans are indeed "managing" the atmosphere. We had better manage it well.

The importance of changing our ways came to me recently as I sat once again on Sally Rock Point, this time with my six-month-old son on my lap. On that chilly winter day, I contemplated the warmer future that will probably confront my son before he reaches old age.

As I watched Daniel's eyes scan the water, my mind filled with images of this corner of the Earth as it might be transformed by the sudden warming resulting from that blanket of greenhouse gases. I saw waves inundating the remains of my parents' abandoned house and washing over the dying salt marshes that had no room to retreat. I imagined beetles and termites devouring the skeletons of the pine forest that once flourished behind the house, but now had shriveled because of drier, hotter summers.

And, strangely, I heard laughter. It was the chuckle of the future residents of Sally Rock Point, laughing incredulously as someone told them a story about an old ice road that once cut across the bay.

A SCENE OF CHANGES

"The world's a scene of changes,
and to be Constant,
in Nature were inconstancy."

—Abraham Cowley,
English poet (1618–1667),
"Inconstancy"

In the turbulent panorama of global climate, weather events occur on all scales: ranging from dust devils, such as the one shown above, in central Washington, to potent thunderstorms. The preceding pages show a storm breaking up in Kenya's Masai Mara. Lightning splits the sky near Norman, Oklahoma (opposite).

Earth's atmosphere presents a paradox: It is in constant flux, yet it is also remarkably predictable. The flux is obvious to anyone who has sat in a field on a blustery day and watched a scudding panorama of clouds, then sun, then a shower, then sun again. Hour by hour, day by day, season by season, weather patterns sweep across the face of the planet—from North Atlantic gales to a line of thunderstorms rumbling across Kansas, from Los Angeles's searing Santa Ana winds to a deep-freeze blizzard in the Alps.

All of this action is ultimately driven by energy from the sun. The great spinning sheath of gases that makes up the atmosphere is constantly being heated and cooled, blended and stirred. Warm air rises; water evaporates, condenses into clouds, and falls as rain or snow. The patterns range in size from the tiny dust devils that stir up leaves as they dance across a field to the globe-spanning jet stream and to hurricanes with the force of a hydrogen bomb.

Yet despite the daily changes in weather, our atmosphere has an overall stability and predictability, which become apparent at larger scales of space and over long spans of time. From satellites and space craft, the atmosphere appears almost serene—that "moist, gleaming membrane of bright blue sky," as Lewis Thomas has described it. When the small gusts and weather fronts and local storms are averaged out, the system begins to show signs of order. Although it is impossible, for example, to predict when and where a particular tornado will strike, it is clear to meteorologists that the prevalence of certain conditions in a swath of the Midwest—dubbed Tornado Alley—makes that area the most likely to be struck. By studying charts of barometric pressure and other data, forecasters there can issue tornado warnings for a particular day.

Climate: A Predictable Pattern

Many such patterns of atmospheric activity are consistent over thousands of years. The parade of the seasons is one of the most fundamental such

One of the most fundamental, recurring rhythms in nature is the cycle of the seasons. These four images of a wooded ravine in Missouri show the typical pattern in temperate climate zones.

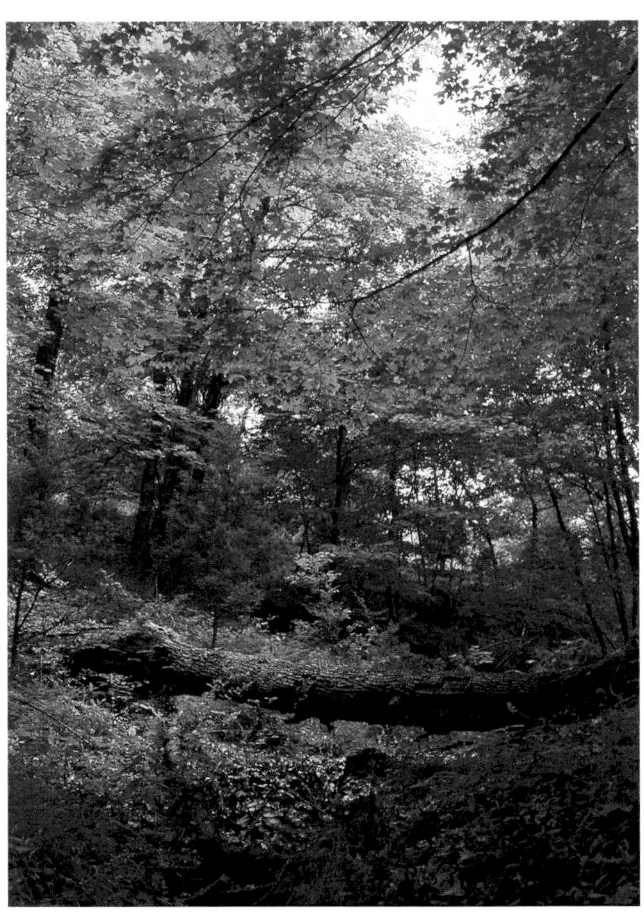

The climate of a region can be changed over millions of years by processes as subtle as the rise of a mountain range, which alters patterns of wind and precipitation. Opposite, the Grand Tetons tower over Teton National Park in Wyoming.

rhythms. We count on the fact that, in the temperate northern hemisphere, April showers *are* usually followed by May flowers. The trade-wind routes that sailors follow today have been in use for centuries, largely unchanged. In the tropics, each day takes on a predictable rhythm, as humidity and heat build tall cumulus clouds throughout the day, until the air can hold no more moisture. Then sudden downpours bring welcome relief. This averaged, smoothed-out, somewhat predictable picture of regional conditions—of general patterns of temperature, moisture, and wind—is called climate.

Climate does change, but very gradually: decade by decade, century by century, millennium by millennium. These changes are caused by factors ranging from slight variations in the Earth's orbit to shifts in ocean currents; from cycles of sunspots, which increase the amount of solar energy reaching Earth, to the gradual growth of a mountain range, which alters wind and moisture patterns.

We all expect weather to change, but it is difficult to consider the changes in climatic conditions that prevail year in and year out. We all have a sense of what the "normal" climate is for our home town, our country, and places we have visited. That sense of what is normal, however, is only a function of our brief experience with weather—a few decades. Human lives are usually too short to allow an individual to observe a fundamental shift in temperature or moisture for a region. When people *think* they have observed such a change—and surveys have shown that many people feel they have noticed "a change in the weather" in their lifetimes—they tend to be wrong. Statistical studies usually show that such subjective impressions most likely reflect a fluke series of warm summers or wet winters or the like.

Indeed, our awareness of the day-to-day changeability of the weather constitutes one of the great impediments to our appreciation of the threat posed by global warming. When a cold snap can cause the temperature outside to plummet 20°, 30°, even 40°F (11°, 17°, or 22°C) in just a few hours, how is someone supposed to get concerned about an eight-degree F (4.5°C) rise in the average global temperature over 100 or so years? As a result, opinions on global warming have in some cases grown cynical; this letter to the editor published in the *San Jose Mercury News* in January 1991 provides a fair example: "Last February I failed to see any stories about the infamous greenhouse effect or global warming during a week of record low temperatures. At the time, I thought you might at least express opinions about the money-grubbing scientists whose defective models had predicted the overheating of our earth." It is only natural to be confused about greenhouse warming when you are shivering through a cold spell.

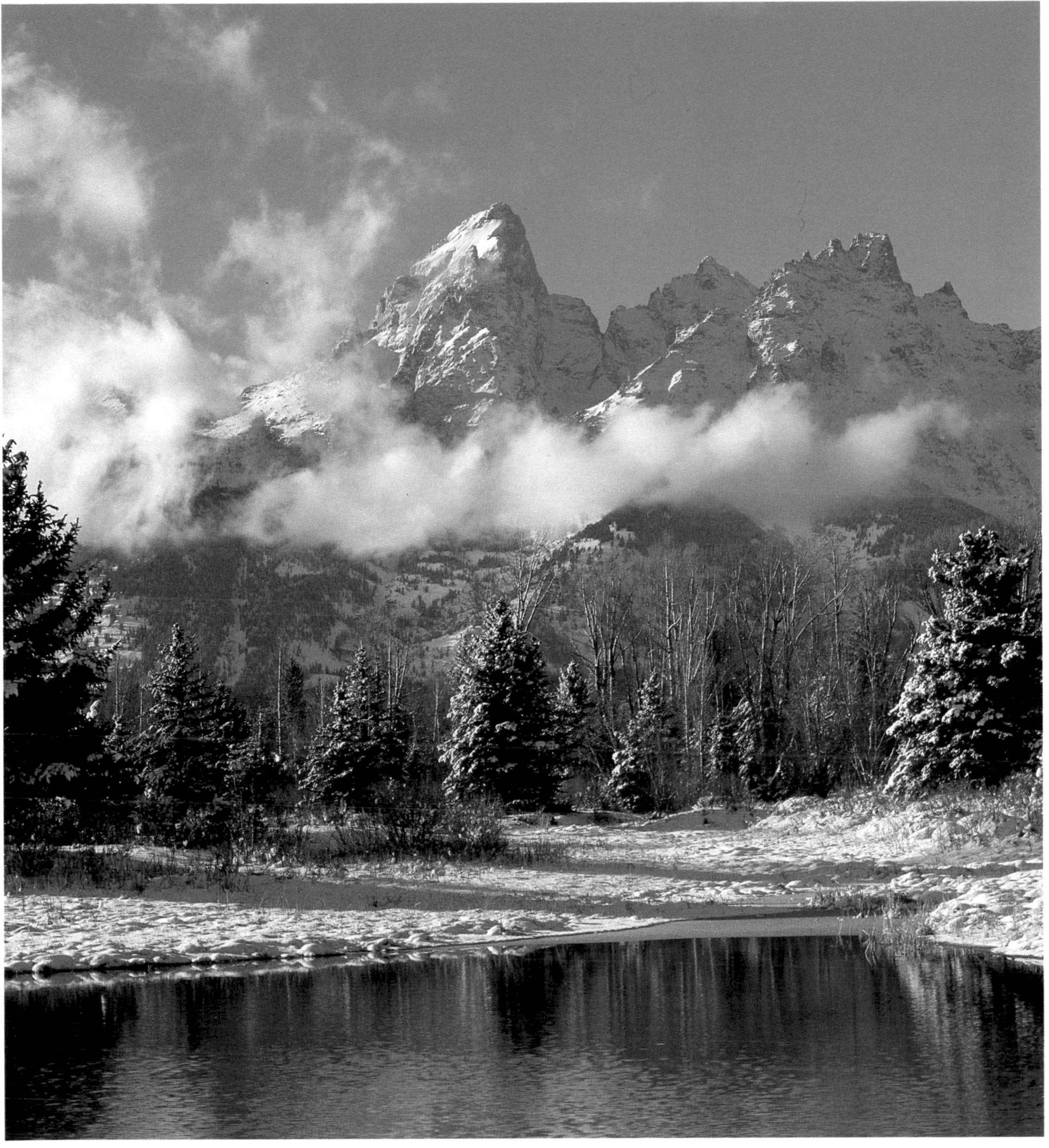

A Break With the Past

Patterns of dramatic changes in climate become apparent, however, when scientists examine long-term records of past conditions. Evidence may be found in a vineyard's century-long log of its harvests, a historian's description of climatic conditions from a bygone age, variations in tree rings, or clues trapped in the layers of a glacier or sedimentary rock. Such records reveal the strong links between climate and life, between climate and human affairs.

Examples of this linkage are everywhere. Take, say, the Sahara, which today is one of the world's most inhospitable spots. Just 9,000 years ago, the Sahara—along with much of the Middle East—was covered with lakes and lush grassland that supported a rich array of life forms. Regular monsoon rains bathed the region. Beneath today's desert sands, fossilized pollen grains indicate the presence just a short time ago of those moisture-loving grasses. In layers of sedimentary rock, formed as dust and eroded soil accumulated at

In this image photographed from the space shuttle, a line of massive rain storms builds over the mountains of northern Madagascar. The daily rhythm of afternoon downpours in the tropics is one of the predictable events that define the region's climate.

Jet streams, high-altitude, high-speed rivers of air, are distinctive features of global climate. Opposite, clouds trace a jet stream over the Nile Valley and the Red Sea.

the bottoms of ancient lakes, the fossilized bones of crocodiles and hippopotamuses can be found. Even the water that is pumped to the surface in the oases scattered through the deserts of the region tells the story. Radiocarbon dating has shown that much of the water there was deposited 10,000 or more years ago. At that time, then, "normal" conditions for the Sahara were temperate and moist.

If you were to turn the geological clock back another 9,000 years, you would find much of the planet locked in an ice age, with glaciers grinding across eleven million square miles of the Northern Hemisphere that are now ice-free. In regions where snow now falls during winter, the snow never melted; layer upon layer of snow compacted into great fields of ice. The sprawling ice caps at the poles influenced wind and moisture patterns all the way to the equator. Where the Bonneville Salt Flats are today, there was a huge shallow lake. Where the Amazon rain forest is today, there appear to have been broad stretches of savanna and small pockets of trees.

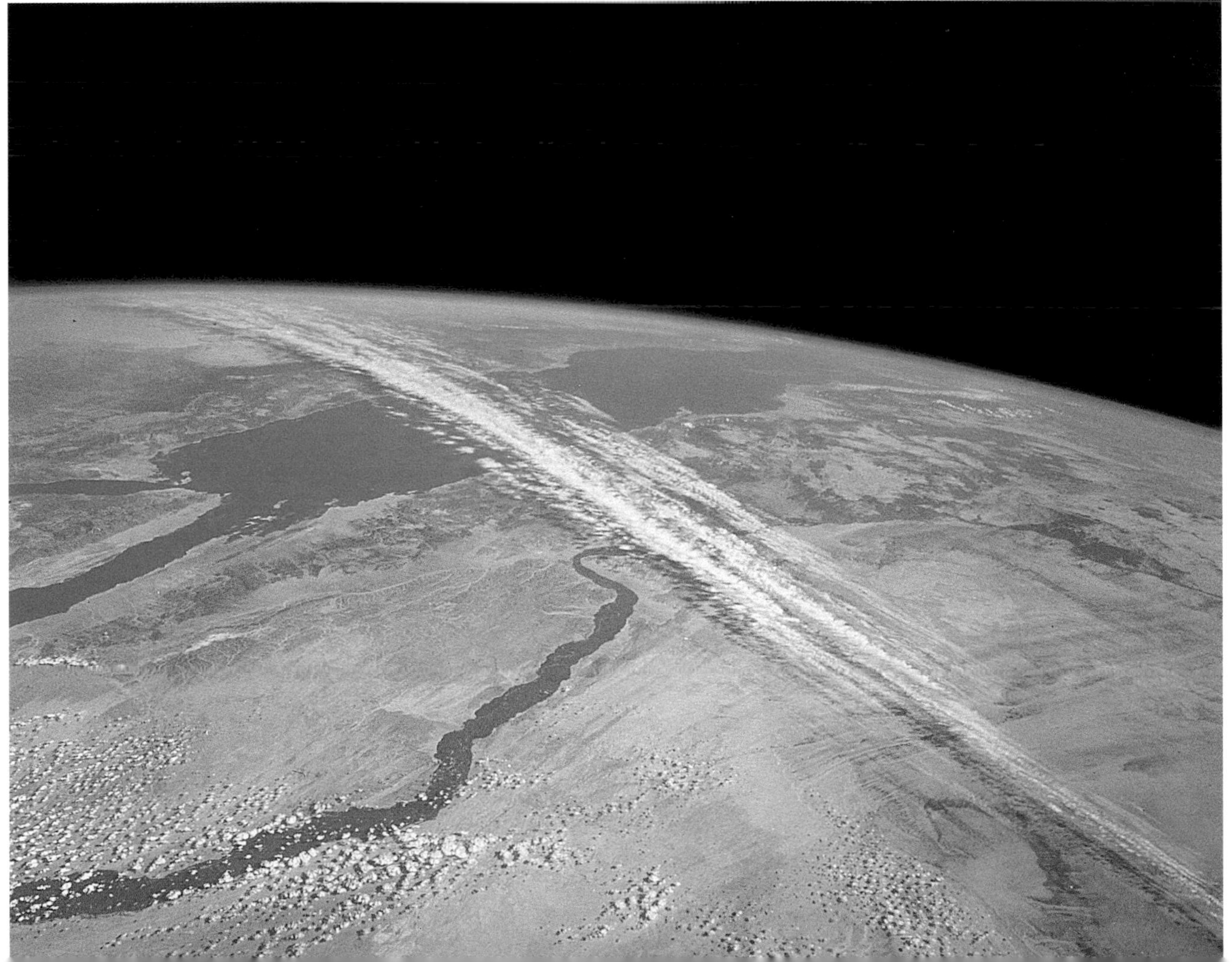

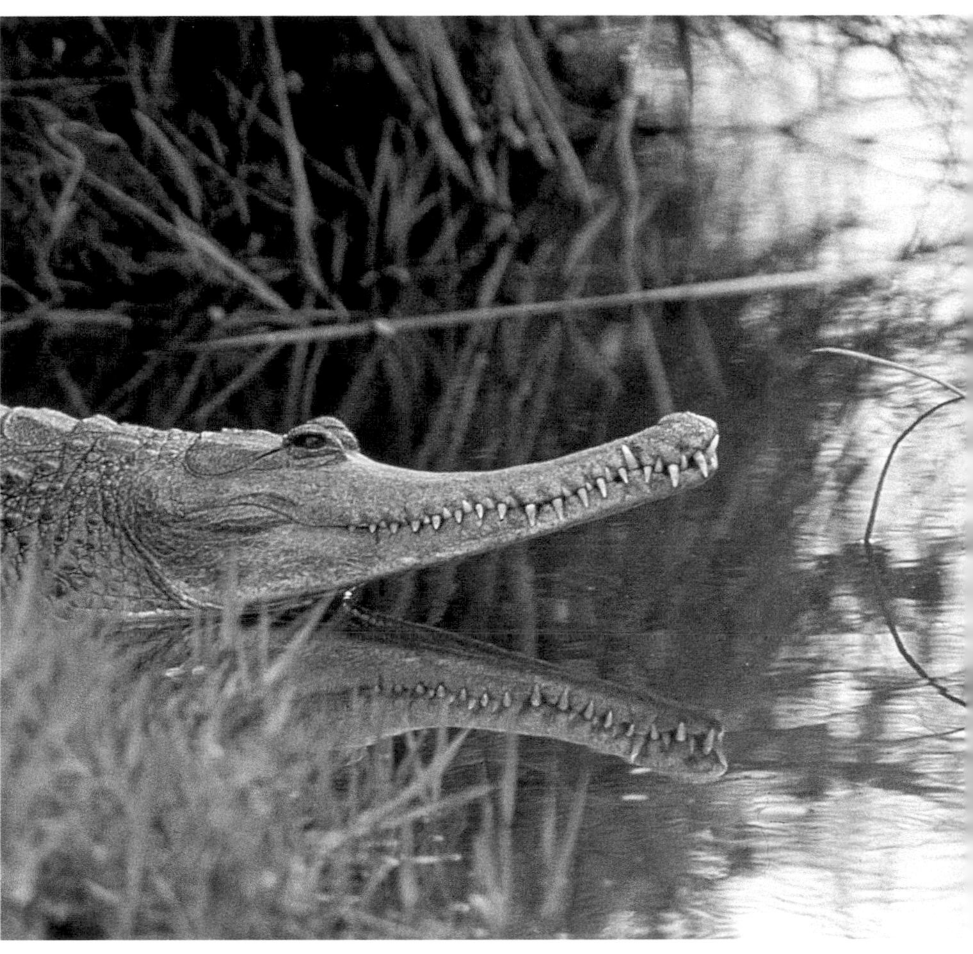

Changes in climate are everywhere accompanied by changes in patterns of life. The Sahara Desert today is dry and harsh (opposite). But 9,000 years ago, the same region was covered with lakes and grassland, and it harbored such water-loving species as the crocodile, shown here on the bank of the Nile River in Egypt (above right).

To get the best perspective on climate changes to come, it is helpful to scan briefly through the Earth's history—to wind the geological clock back to the point when the atmosphere first formed, and then when it attained the composition it has today.

On the newborn planet, some four billion years ago, volcanoes disgorged billions of tons of water vapor, sulfur, carbon dioxide, methane, ammonia, and other materials, creating a shroud of gases. When the surface of the planet cooled below the boiling point of water, water vapor condensed and fell from the skies in a steady rain. There is ample evidence that just 200 million years after Earth formed, it developed one of its two most distinctive features: oceans of water. Just another few hundred million years later, Earth's other distinctive feature, life, appeared.

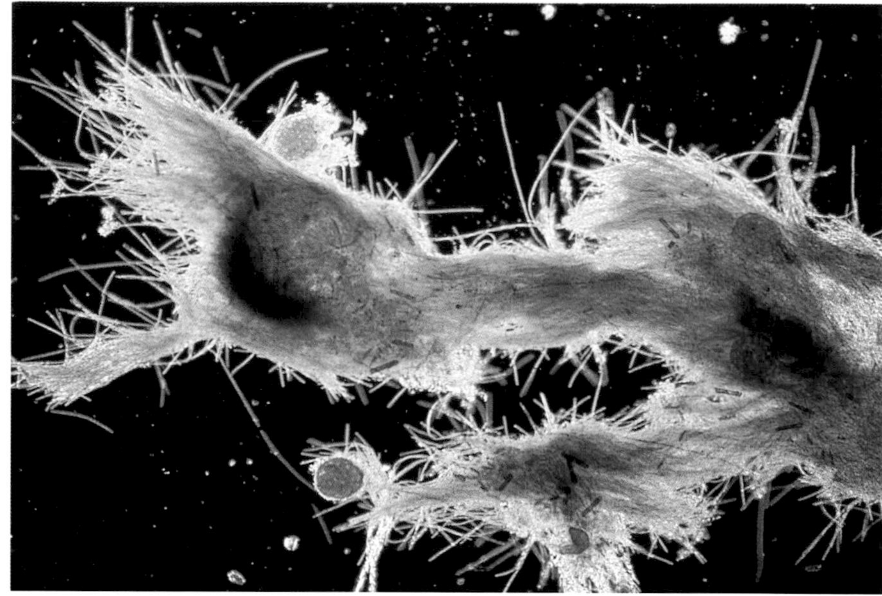

Blue-green algae, shown here in a photomicrograph, were among the first creatures able to turn sunlight and water into food. Starting over 2 billion years ago, such organisms began to fill the atmosphere with oxygen.

From the remote moment when a stew of organic, or carbon-based, molecules were somehow organized into strands of genetic material, the fate of the planet was forever changed. Thenceforward, the atmosphere and the oceans and the substance of the planet itself would be intricately interrelated with the ever more complex life forms that came to inhabit it: first colonies of bacteria, then algae, then multicelled animals, then complex green plants, and eventually human beings and their machines.

The main link between this biosphere and the atmosphere was photosynthesis, a biochemical process that allows certain microbes to convert sunlight, carbon dioxide, and water into food. Photosynthesis is the process that produces redwoods, apples, and roses. It was the world's first solar-powered engine. This chemical reaction indirectly created the reserves of oil and coal upon which modern industrialized society feeds.

Scientists estimate that the earliest photosynthesizing microbes bloomed in the sea some 2.8 to 3.5 billion years ago. At that time, the atmosphere above the oceans contained far more carbon dioxide—at a concentration perhaps a thousand times greater than now. In the air and water, oxygen was present in only the tiniest traces, and it was toxic to life forms.

At this time, life exerted its first dramatic influence on the planet. Photosynthesis releases oxygen as a byproduct. The first photosynthesizers, like other early life forms, still could not tolerate free oxygen—it was truly a toxic

Stromatolites, lumpy reefs produced by blue-green algae, still exist in a few places, such as Shark Bay, Western Australia. Fossilized stromatolites have been found dating back more than 3 billion years.

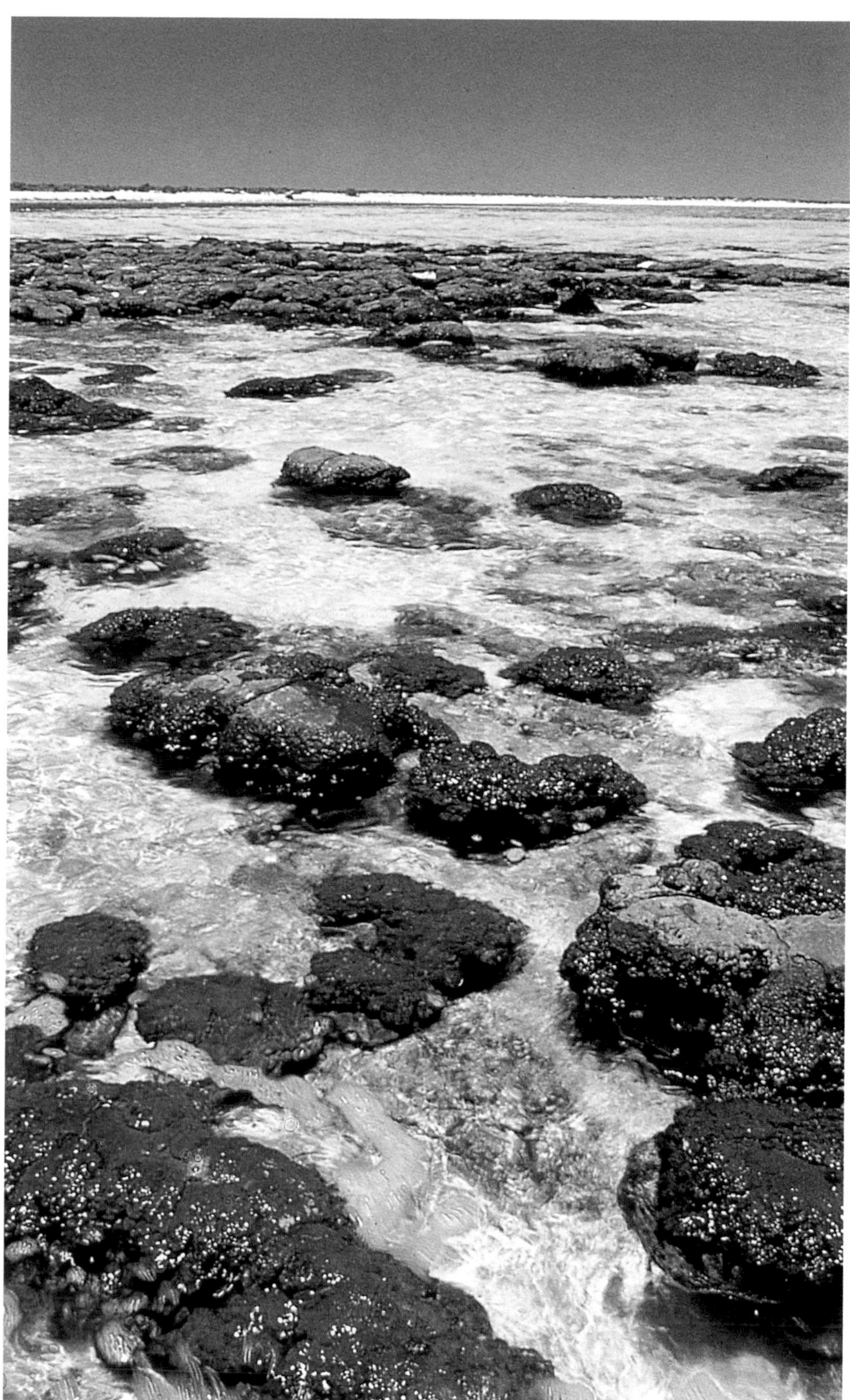

A diorama at the American Museum of Natural History depicts life in the oceans of the Ordovician Period, some 500 million years ago. By this time, the planet had an atmosphere rich in oxygen.

waste. But as photosynthesizing life continued to evolve, natural selection produced organisms that were able to thrive in an oxygen-rich environment. These innovators soon dominated other forms of life. As they spread, the free oxygen they produced diffused into the atmosphere.

Creation of the Ozone Layer

A minor side show took place at the time that would prove to have important consequences later on. Some of the oxygen rose to the highest regions of the planet's atmosphere, where ultraviolet radiation from the sun caused a reaction that formed fragile triplets of oxygen atoms—O_3 or ozone. A diaphanous veil of this ozone developed, a layer that effectively absorbed the

By the Carboniferous Period, 325 million years ago, rich fern forests, depicted here in a museum diorama, covered large areas of Earth's land surface. Much of the planet's stores of coal are the fossilized remains of these forests.

destructive energy of the sun's ultraviolet radiation. Ultraviolet radiation can easily shatter genetic material—thus the formation of the ozone layer probably enabled life to leave the protection of the sea. Without that shield, it is doubtful that plants, and later, the first animals, would ever have crept onto dry land. (And now that ozone layer is being sapped by synthetic chemicals manufactured by one of the species that resulted from life's first forays ashore.)

The end result? By about 500 million years ago, the atmosphere, influenced by biological processes, had been gradually but dramatically transformed: from a primordial envelope of nitrogen and carbon dioxide to a mixture approaching 78 percent nitrogen, 21 percent oxygen, and a trace of

carbon dioxide and other gases. Oxygen, carbon, and nitrogen atoms were passed from air to organism to earth and water then back again. For example, consider a carbon atom, C, in a CO_2 molecule. That C might circulate in the air for years, then dissolve in the ocean, be taken up by a microbe through photosynthesis and incorporated into a shell of calcium carbonate ($CaCO_3$). When the organism died, the shell would drop to the ocean bottom, be transformed into limestone, then many millions of years later be disgorged back into the atmosphere as carbon dioxide when that now-ancient rock was consumed by geothermal heat and exhaled by a volcano.

For perhaps 500 million years, then, the Earth—cloaked in an insulating atmosphere capped by a protective ozone shield—has maintained a remarkably equable climate and atmospheric chemistry. The amounts of the predominant gases, nitrogen and oxygen, have stayed virtually constant. The global average temperature has never dropped below freezing and never risen much above the hottest readings found in today's deserts.

Mass Extinctions

The system has taken some remarkable abuse, such as occasional direct hits by massive meteorites or asteroids—including one collision that is thought by many scientists to have ended the age of dinosaurs 65 million years ago. Life on the planet has suffered other large-scale assaults. As continents formed, drifted together, and split apart, changes in ocean currents, volcanic activity, ice sheets, and other factors may have contributed to periodic massive die-offs of species. The fossil record is punctuated by five such mass extinctions.

A crisis for one species, however, is an opportunity for another. The biosphere has always bounced back, with plants and animals rapidly filling niches vacated by organisms that went extinct. The disappearance of the dinosaurs, for instance, was quickly followed by an explosion of evolution in mammals.

Interestingly, the dynasty of the dinosaurs, from 220 million years ago to the time of their demise, was one of the last long periods of consistently warm, wet weather in the planet's history. There is quite a bit of evidence showing that, 100 million years ago, the world was more uniformly warm than today, with no significant glaciation, even at the poles. At that time, great masses of vegetation lived and then died in what is now Antarctica.

Soon after the dinosaurs disappeared, something changed. The global temperature began a slow slide toward cooler conditions. Two million years ago, for reasons that are not yet adequately explained, the planet descended

Sixty-five million years ago, the dramatic disappearance of dinosaurs was quickly followed by an explosion of evolution in mammals. At right are two Trachodon dinosaurs. By the Pleistocene, 1.8 million years ago, mammals of myriad types inhabited the continents. Below is a diorama, based on fossils found in North America, showing a saber-tooth tiger.

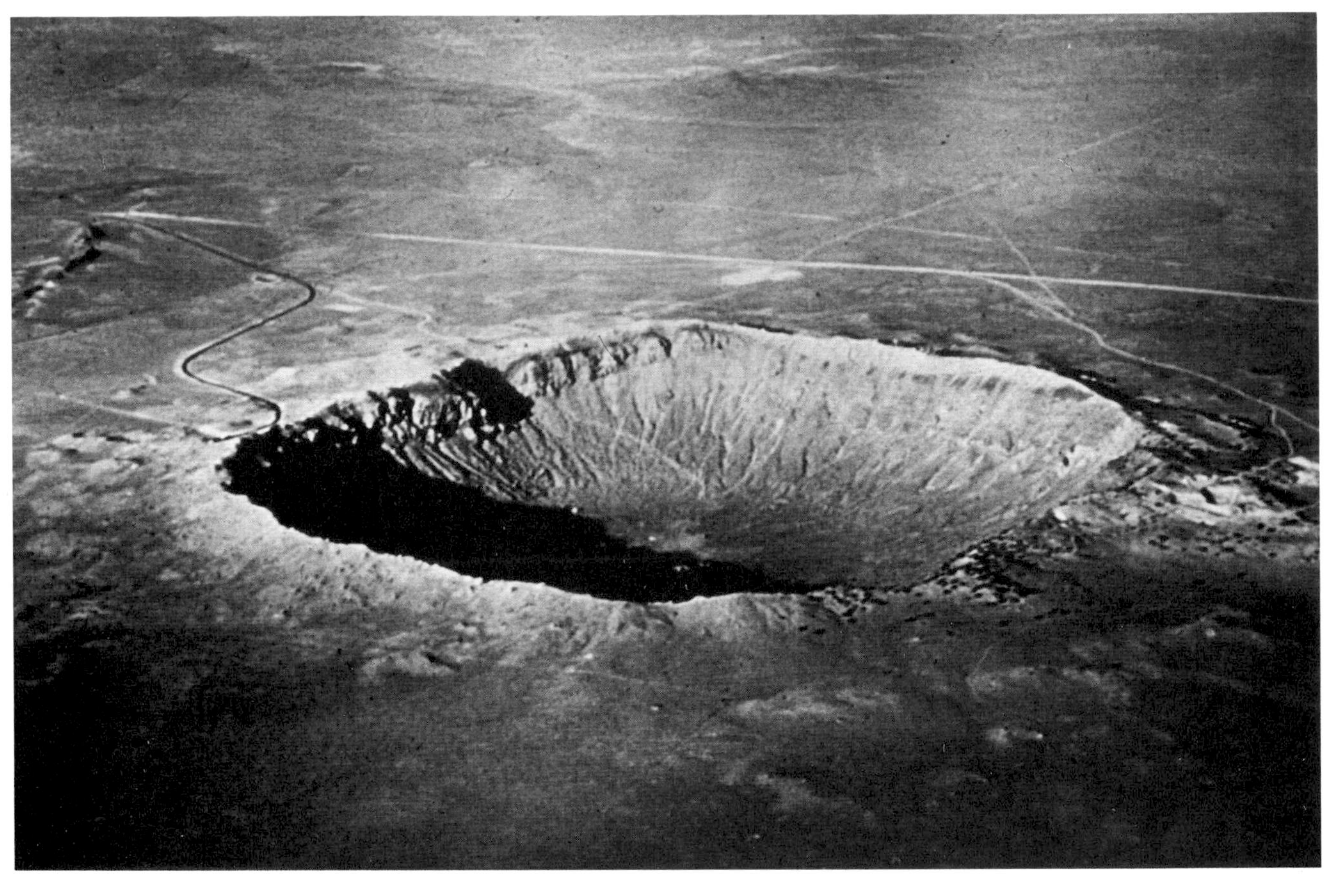

Earth has periodically been struck by massive meteorites, some of which are thought to have caused mass extinctions. This crater in Arizona, just under one mile in diameter, is thought to have been produced by a collision with a 650-foot-wide meteorite.

into an epoch of ice—a regular cycle of long ice ages and brief respites, called interglacials. (In the next few thousand years, the current interglacial is expected to end.) Roughly every 100,000 years—in a cycle believed to be dictated by changes in Earth's orbit and rotation—the ice sheets at the poles have crept toward the equator and then retreated. As much as one third of all the Earth's land area has been covered with ice at the peaks of these glacial periods.

Survival of the Fittest

Ever since the beginning of the present epoch, called the Pleistocene, 1.8 million years ago, all successful forms of terrestrial life have had to shift or adapt in response to the advance and retreat of the ice. This has been an unrelenting cycle of change, allowing little time for the biosphere to sit idly. For the last 160,000 years, an especially detailed record of climate has been deduced from fossils, the chemistry of ancient ice, and other evidence; and in this period the global temperature has steadily risen, fallen, and risen again.

In North America, for example, studies of fossilized pollen have shown the resulting advance and retreat of maple and oak forests, which need relatively temperate conditions, and a more northerly band of spruce and other coniferous trees, adapted to colder conditions. The impact of the ice ages is felt all the way to the equator and extends into the seas as well. Ancient layers of coral beneath today's reefs show how sea levels rose and fell hundreds of feet as more or less water was locked up in glaciers.

One particularly adaptable and innovative species can trace much of its lineage within this age of rhythmic climate change. That species, of course, is *Homo sapiens*. Virtually the entire known span of human history takes place in the Pleistocene. The first evidence of hominid use of fire—some charred bits of antelope bone from a cave near Pretoria, South Africa—dates from 1.2 million years ago. Much of the great expansion of the human species over the face of the globe has taken place in just the past 30,000 years or so, since the onset of the last ice age.

All of modern civilization has blossomed in a short respite from the overarching era of cold—the most recent interglacial, which geologists call the Holocene. Until 10,000 years ago most of the heart of western Europe, from the British Isles east through Germany, was bleak tundra. Only after a centuries-long warming trend did European populations grow and agriculture develop. The Sumerians flourished in what is now southern Iraq starting only 6,000 years ago. Five thousand years ago, an especially warm, humid period may have set the stage for the first flowering of Chinese culture.

Global Average Temperature

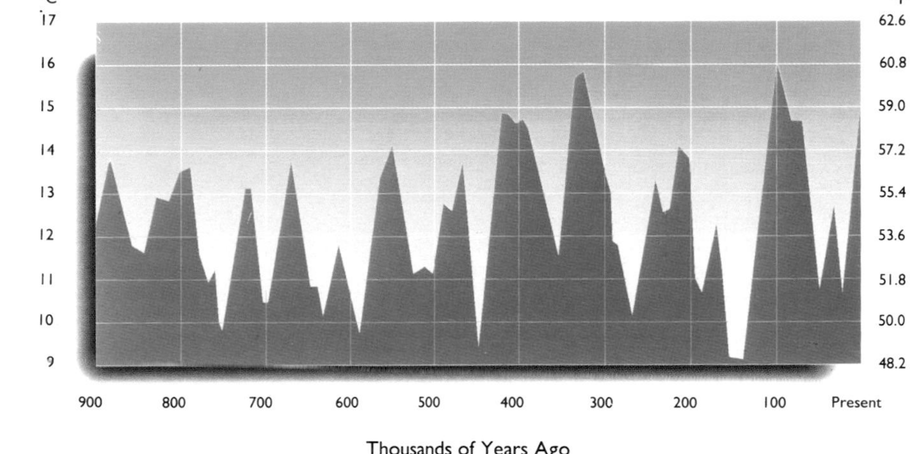

For hundreds of thousands of years, Earth has been in a cycle of ice ages and warm intervals. Most of these climate changes took place over thousands of years. A much faster temperature shift from the growing greenhouse effect may occur in the next century.

Even within the relative warmth of the Holocene, little flutters of cold and warmth and drought have forced human societies to shift. There are instances in which a change of just a couple of degrees in the average temperature caused a major shift in a society (an important lesson to keep in mind as we approach an expected warming of as much as eight degrees). A warming trend in Europe from A.D.950 to 1250—sometimes called the Medieval Optimum—allowed Vikings to colonize previously inhospitable spots such as Iceland and southern Greenland. Greenland, in fact, never was very green, but Eric the Red gave it that name to entice more settlers to migrate there. At its peak, the Greenland settlement had 280 farms and a population of 3,000. At around the same time, dozens of vineyards flourished in Britain—so many that France wanted to limit imports from its island neighbor.

Gradually, though, the climate of the northern hemisphere cooled. Most of Britain's vineyards were put out of business. Greenland became increasingly locked in sea ice. By 1492, Pope Alexander VI had noted reports that Greenland was almost unreachable. "Shipping is very infrequent because of the extensive freezing of the water—no ship having put into shore, it is believed, for eighty years," he wrote. The settlement eventually died out.

From 1500 to 1850, much of Europe, North America, and other parts of the globe experienced what has been called the Little Ice Age. Many regions

During a prolonged warm period, from A.D. 950 to 1250, Vikings were able to colonize Iceland and Greenland. At left is a medieval church in Iceland; note the sod roof. Once the cold weather returned, the Greenland settlement was abandoned. Today, Greenland is populated again. Opposite is a settlement called Ammassalik.

The past century has seen a dramatic retreat of glaciers in such places as New Zealand and the Alps. Above left, a glacier loomed over the French village of Argentière when this engraving was made c. 1850–60. By 1966, when the photograph above right was taken, the river of ice had pulled well back into the Alps.

had sharply colder winters, whose effects are documented in records from French vineyards and in Dutch accounts of disruptions in canal travel because of thick ice. The cold also affected some major wars of the time, creating harsh conditions for American troops at Valley Forge and for Napoleon on his ill-fated march into Russia. Glaciers advanced dramatically in the Alps and in parts of New Zealand. The Thames River in London began to freeze regularly, resulting in the advent, in the winter of 1607, of "Frost Fairs," in which a small tent city sprung up on the river, offering amusements that included ice bowling. In 1662, the sport of ice skating was introduced from the Netherlands at such a fair. The last Frost Fair was held in 1814. Since then, warmer conditions have kept the river from freezing completely.

Past climate changes have strongly influenced human affairs. From 1607 to 1814, "frost fairs" were held on London's Thames River, which routinely froze each winter. This woodcut depicts the fair of the bitter winter of 1683–84. Today the Thames is free of ice.

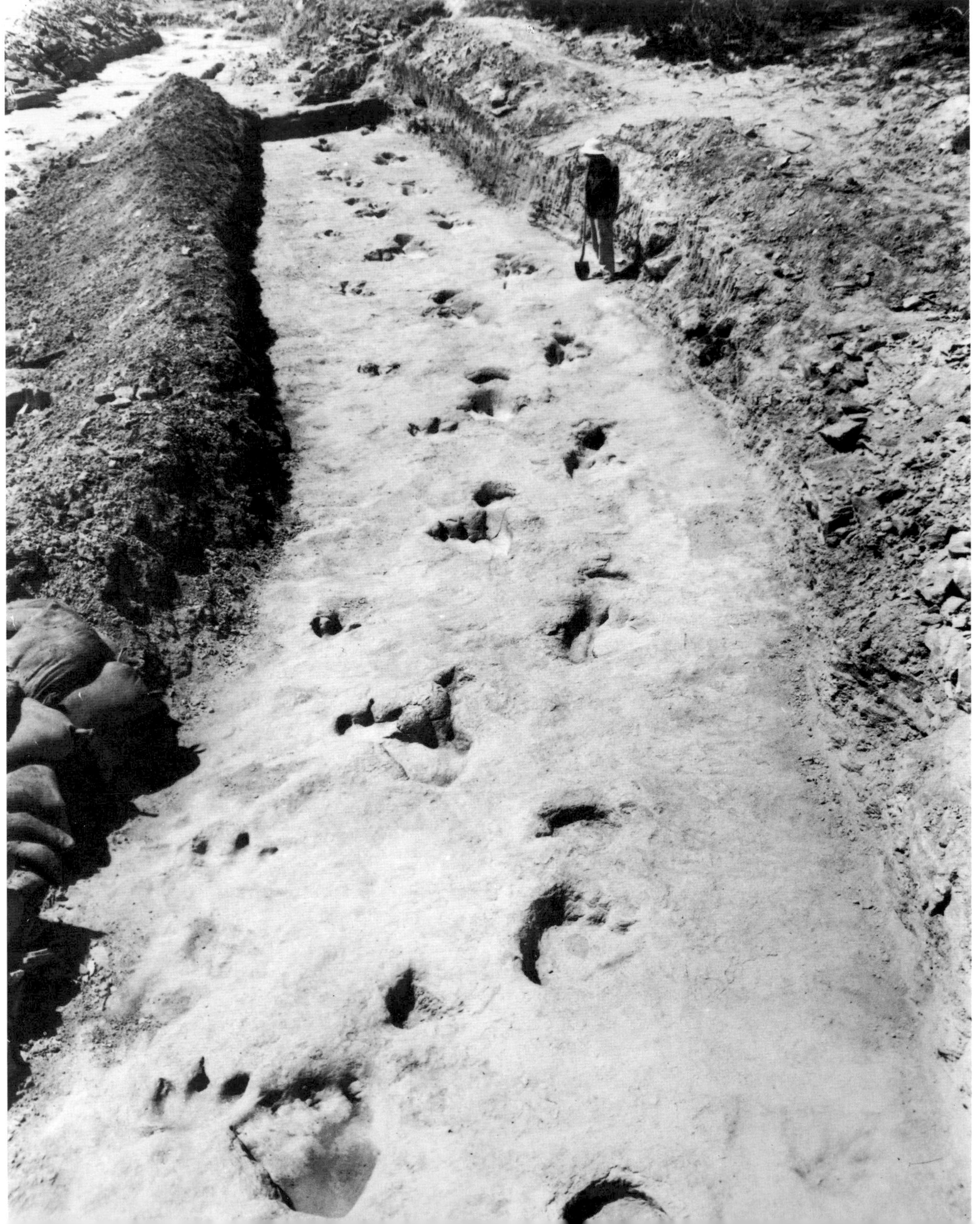

Bracing for Change

Human beings and the rest of the inhabitants of planet Earth may now have to brace for a new, and much more drastic, period of change. Perhaps 2 billion years ago, the fate of the planet was forever altered by living things, as photosynthesis flooded the atmosphere with oxygen. Now, a life form is influencing Earth's fate once again, as the explosive expansion of human populations and industry dumps tens of billions of tons of carbon dioxide and other heat-trapping gases into the air.

Humans, at least, have proved able to adapt themselves to continual shifts in climate—when they have been gradual. The new era of global warming we may face, however, is predicted to occur much more rapidly than any change in the last 10,000 years.

Perhaps earth scientists of the future will name this new post-Holocene period for its causative element—for us. We are entering an age that might someday be referred to as, say, the Anthrocene. After all, it is a geological age of our own making. The challenge now is to find a way to act that will make geologists of the future look upon this age as a remarkable time, a time in which a species began to take into account the long-term impact of its actions. The alternative may be to leave a legacy of irresponsibility and neglect of the biosphere that could eventually manifest itself in the fossil record as just one more mass extinction—like the record of bones and footprints left behind by the dinosaurs.

In changing the composition of the atmosphere, humans may produce a very rapid climate shift. The result may be a new mass extinction, such as the one that ended the era of the dinosaurs. All that remains from their reign are fossilized bones and footprints, such as these at Glen Rose, Texas.

THE GLOBAL GREENHOUSE

"We are evaporating our coal mines into the air."

—Svante Arrhenius,
Swedish chemist (1859–1927),
in an 1896 essay

In the record-hot summer of 1988, which highlighted the vulnerability of natural ecosystems, massive forest fires took place from France to Wyoming. Opposite, bison move through a burning landscape in Yellowstone National Park. Above, a black bear in Yellowstone seeks refuge in a scorched pine tree.

On the preceding pages, trillium blossoms pop up through the snow in Michigan, signaling an early spring. Such sights may be commonplace in coming decades as climate shifts disturb the life cycles of plants.

The first winter of the 1990s was a warm one in my town, Brooklyn, New York. As February began, crocuses and even a few tulips popped their green heads up through the garden soil, fooled by weeks of freak warmth in which New Yorkers donned T-shirts and flocked to parks and beaches.

The warm winter followed a year that set a new record for the warmest global mean temperature, 59.8°F (15.4°C), in the 110 years since such figures had been calculated. In setting that record, 1990 continued a trend begun in the 1980s; at the time, the seven warmest years on record were, from hottest to coolest: 1990, 1988, 1981, 1987, 1983, 1980, and 1989.

This series of hot years caps a hundred-year trend of slow warming; as a result, some normally circumspect atmospheric scientists have gone out on a limb and declared that we are seeing a signal that human activities are exacerbating the greenhouse effect and warming the planet. Keep in mind that atmospheric science is not a field that attracts high-profile types, eager for the spotlight. These are researchers who would much prefer to sit and tinker with their computer models than address congressional hearings. Nonetheless, out they have come—braving the scorn of skeptics.

Through the 1970s and early 1980s, scientific papers on the greenhouse effect had been published with little fanfare. There had been a few congressional hearings on the subject, but no one took much notice. The obscurity of the issue ended as the United States began to wilt during the scorching, endless summer of 1988. That was the summer that saw the forests of Yellowstone National Park and the forests of France go up in smoke, the summer that ruined crops from Canada to China. No one could prove that that particular heat wave—or any single heat wave—was caused by the buildup of carbon dioxide, but the lengthening list of record-hot years in the 1980s was becoming harder to ascribe to any other cause. As James Hansen, who leads a greenhouse research effort at the Goddard Institute for Space Studies, put it,

Global Average Temperature

A graph of global mean temperatures from 1880 to 1987 shows a pronounced warming trend. The total rise has approached 0.9°F (0.5°C).

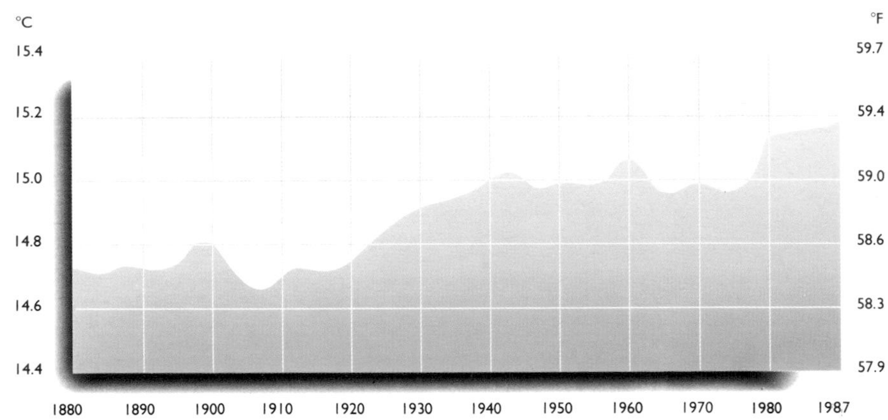

"It is time to stop waffling so much and say that the evidence is pretty strong that the greenhouse effect is here."

Since then, Hansen and a growing chorus of atmospheric specialists have not changed their views. They stress that neither the drought and heat of the summer of 1988, nor the mild winter of 1990–91, nor any other single climatic anomaly can be linked directly to rising levels of heat-trapping gases. But the increasing frequency of warm summers and winters—particularly the rising temperature of the planet as a whole, which is the result of averaging hundreds of separate thermometer readings—is consistent with the theory that humans are adding to the greenhouse effect. As Hansen explained, "Seasonal weather is still a crap shoot, but the global warming is loading the dice."

Starley Thompson, a climate modeler at the National Center for Atmospheric Research, vividly described the growing acceptance of the theory of global warming. He said, "There are always going to be a few hold-outs—'Flat Earthers.' Apart from those, though, I don't think it'll be too long before you see broad agreement on this. A clincher will be this continual occurrence of years that are hotter than any other year in the historical record. Right now it's like a plant is peeking up above the weeds. The plant has to get tall enough to grow out of the weeds. If it continues on this way, definitely, within a decade, all reasonable people will have to sit up and take notice."

The Invisible Problem

Generally speaking, we tend to worry about environmental problems only if they are tangible or visible. Water pollution is an industrial sewer spewing foamy toxins into a greasy lake; air pollution is the sooty blast from a bus's exhaust pipe, or the cloud of yellow smoke rising from a power plant. The crisis of global warming, however, focuses on some of the rarest gases in the atmosphere—gases measured in parts per million, and all of them invisible.

At the center of all the fuss is a gas that we all know from grade school as one of the basic substances of life. We exhale it and plants inhale it. Dry ice is made of it. It can snuff out a match. It is the bubbles in beer. How can carbon dioxide, such a seemingly innocuous compound—just a couple of oxygen atoms linked to a carbon atom—be the cause of such a big problem? How can CO_2 and these other gases act like a stifling greenhouse?

How the Greenhouse Effect Works

On one of the few frigid afternoons of February 1991, my wife and I walked with our infant over to the Brooklyn Botanic Garden. To take away the chill, we thought we would head to the forest exhibits, each enclosed in a lofty, domelike glass greenhouse. Just inside the entrance to the grounds, we passed a big boulder sitting inconspicuously near some bushes. I had passed it a hundred times before, but this time I took a closer look. The boulder, about six feet tall and rounded like a pear, had a smooth, polished spot where countless human rear ends had found a perch. At around eye level above that seat, there was a little bronze plaque embedded in the stone. It read, "Boulder of diabase. Geological age, Triassic. Transported by continental glacier during the Ice Age from Palisades, between Hoboken and Englewood." Here was yet another reminder of the dynamic, ever-changing face of Earth. Plucked from a cliff of 200-million-year-old rock as a glacier scuffed its way across North America 20,000 years ago, this boulder was carried along like a pebble caught in the tread of a child's sneaker, then dropped as the ice melted back to the north.

We made our way to the greenhouses, where we began in the temperate forest, a replica of Mediterranean conditions where vents in the glass roof keep the chamber cool and dry. The pavilion has beautifully landscaped slopes covered in silvery shrubs, mostly hardy varieties that tolerate dryness. Then we opened the doors leading into the tropical rain forest exhibit, leaving Greece behind in an instant and arriving in the Amazon. It was early February in New York, yet suddenly we were immersed in a steaming hothouse, rich with the scents of citrus and coffee blossoms, moist earth and fungi. The sun shone as brightly outside the dripping panes of glass as in, but we were sweltering in our parkas, while people outside were thankful for theirs.

Tropical plants thrive under a greenhouse roof in the rain forest exhibit at the Brooklyn Botanic Garden. Much as the glass roof of a greenhouse traps heat, carbon dioxide and several other gases trap heat energy in the atmosphere, giving Earth its equable climate.

Carbon dioxide and the other so-called greenhouse gases act as a heat trap in much the same way that the glass panes of that Brooklyn greenhouse do.

The atmosphere was first compared to a "glass vessel" in 1827 by the French mathematician Jean-Baptiste-Joseph Fourier. He recognized that the air circulating around the planet lets in sunlight—as a glass roof does—but prevents some of the resulting warmth from leaving. In the 1850s, the British physicist James Tyndall took things further and tried to measure the heat-trapping properties of various components of the atmosphere. Surprisingly, he discovered that the two most common gases, nitrogen and oxygen, have

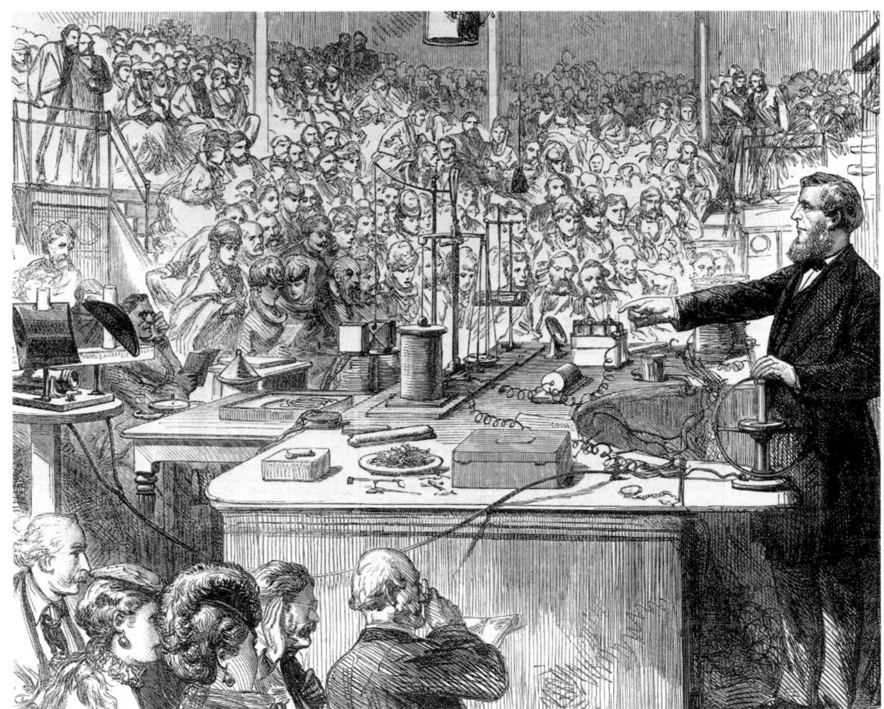

The British physicist James Tyndall was the first scientist to measure the heat-trapping properties of different atmospheric gases. Here he gives a lecture at the Royal Institute of London.

no heat-trapping ability; 99 percent of the atmosphere has no insulating properties at all. It is up to a few trace gases—water vapor, carbon dioxide, methane, and the rest—to keep the planet cozy.

The Physics of the Greenhouse

Since Tyndall's time, the process by which these trace gases keep the planet warm has become clear. Experts still engage in rancorous debates over how much, when, and where the planet may warm as these gases increase, but they agree on the basic physics. Most of the sun's energy travels to Earth as visible light. The sunlight enters the atmosphere and warms things up—particularly things that are dark in color and thus absorb a lot of light, things such as plants, soil, and the oceans.

Surfaces warmed by the sun then begin to shed that accumulated energy in a different form, as heat, which is simply energy radiating at an invisible part of the spectrum—called the infrared. Think of a rock that is tossed into a campfire. It is heated by the flames, then, long after the fire is out, you can still feel heat radiating from the rock. That "heat" is infrared radiation.

The greenhouse gases act like an insulating blanket. Without the green-

How the greenhouse effect works: Sunlight enters the atmosphere; some of it (white arrows) is reflected by clouds, and some (pale blue band at center) passes through the atmosphere and strikes the oceans and the land, warming the planet's surface. The land and water radiate some of that energy back out toward space in the form of lower-energy infrared radiation (orange arrows), but this time some of the radiation is absorbed, raising the atmosphere's temperature.

Various gases—water vapor, carbon dioxide, chlorofluorocarbons, methane, and nitrous oxide—act like a blanket, trapping more of the infrared energy before it can escape and further warming the atmosphere. As amounts of these gases grow, the effect may turn up the global thermostat.

Greenhouse gases are increasing as a result of diverse human activities, represented in scenes from left to right: the dumping of refuse in landfills; agriculture, including rice cultivation and cattle breeding; automobiles; power plants; the burning of forests and grasslands; and the energy demands of cities and towns.

Chlorofluorocarbons or CFCs (peach bands) escape from refrigerators, some foam insulation, home and automobile air conditioners, and some industrial processes. Methane (purple bands) is released by bacteria in landfills, rice paddies, and the guts of cattle, as well as by fossil fuel production and the burning of forests. Nitrous oxide (gray bands) is released by fertilizers; other sources are not fully understood. Carbon dioxide (green bands) comes from automobiles; the burning of forests; and the use of fossil fuels to generate heat, electricity, and industrial power.

house effect, heat energy would quickly radiate back into space from the Earth's surface, leaving the planet with a temperature close to 0°F (−18°C), instead of the present comfortable average of 59°F (15°C).

Just as a tuning fork with prongs of a certain length starts to hum when placed in the presence of sound at just the right pitch, molecules with a certain shape start to vibrate when exposed to energy of a particular wavelength. Molecules of carbon dioxide, methane, water vapor, and the other greenhouse gases are not "tuned" to absorb energy transmitted as visible light—such light passes right through them, as it does through a transparent window pane. Molecules of the greenhouse gases, however, are exquisitely sensitive to infrared energy. They absorb and re-emit it, warming the atmosphere close to Earth.

Even though the greenhouse gases exist as only a trace, they exert a powerful influence on the temperature of the planet. Water vapor, for example, constitutes just one percent of the atmosphere, carbon dioxide a mere .035 percent. Yet without these tiny proportions of greenhouse gases, Earth would be a deep-frozen snowball, some 59°F (33°C) colder.

The scarcity of these greenhouse gases makes clear their potency. A

Other planets in the solar system illustrate the significance of the greenhouse effect. Venus, below right, has a dense atmosphere that is mainly carbon dioxide. Its sterile surface, depicted below in a computer-enhanced radar image, sizzles at 890°F (477°C).

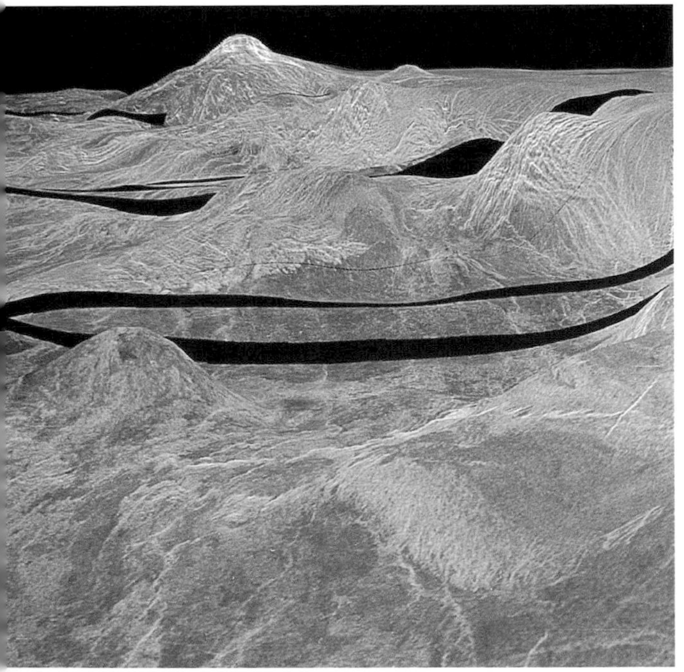

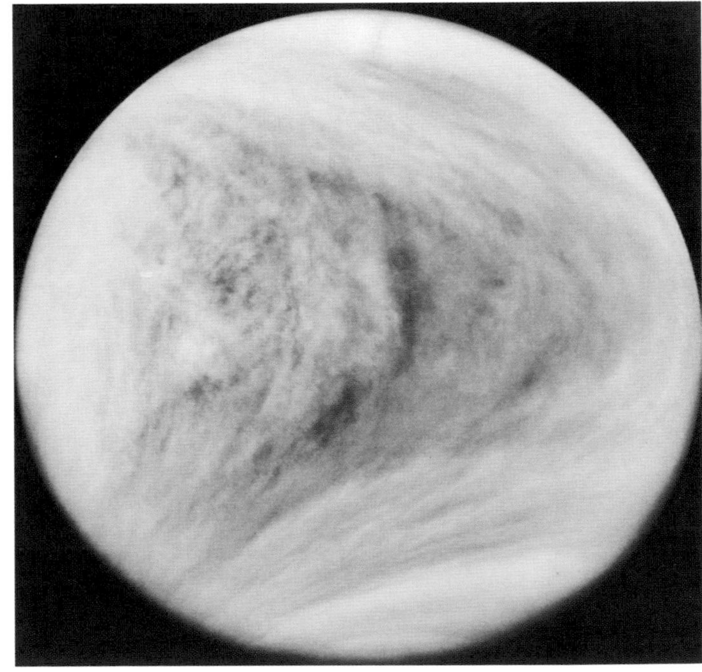

small change in their concentrations can cause a big change in the way the atmosphere behaves. And now human activities are threatening, in a few decades, to double the amounts of several of them.

There is no better way to appreciate the importance of greenhouse gases in determining a planet's climate than to look at three convenient experiments: Venus, the second planet from the sun; Mars, the fourth; and Earth, in between them. Venus has a dense atmosphere that is thick with carbon dioxide and other gases that trap heat. As a result, the planet has a runaway greenhouse effect, producing a surface temperature of 890°F (477°C), hot enough to melt tin and lead. The atmosphere on Mars is mostly carbon dioxide, as well, but is very thin. And Mars has barely a whiff of water vapor in its atmosphere to help trap the sun's heat. With little greenhouse warming, Mars has a mean temperature of 53°F (47°C) below zero, colder than that of Antarctica.

Earth, literally and figuratively, lies between these two extremes. The planet is cloaked in an atmosphere with sufficient quantities of water vapor and carbon dioxide to retain some of the heat of the sun, but not so much that the temperature exceeds the range in which life can flourish. Planetary

Mars, below right, has a thin atmosphere that has little greenhouse effect. Its average surface temperature is colder than that of Antarctica. The surface is stark and still (below). What water exists on the planet is in a perpetual deep freeze.

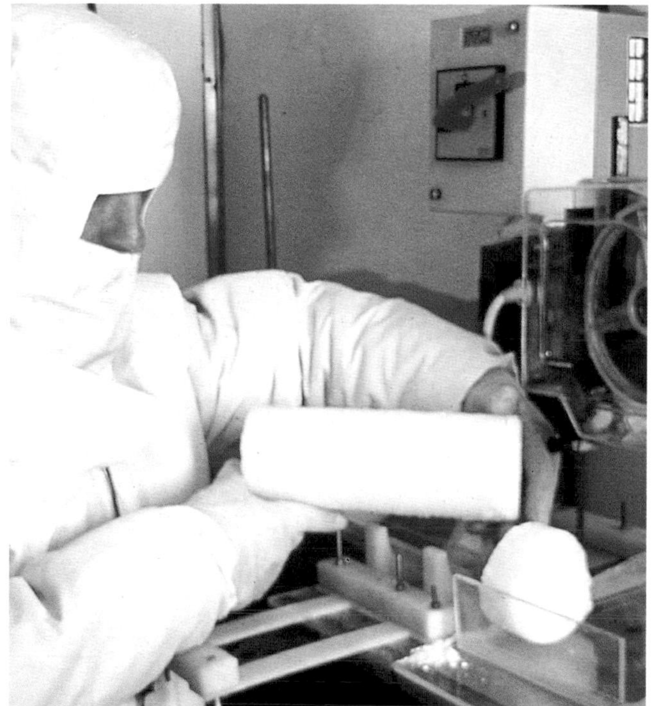

Scientists extract an ice core in Antarctica (opposite). Above, a British researcher cuts a section of such a core. At right above is a microscopic view of a cross section through the ice. Bubbles contain ancient air. The composition of the ice indirectly indicates global temperature.

scientists have referred to the situation of Venus, Mars, and Earth as the Goldilocks phenomenon, in reference to the bowls of porridge left behind by the three bears: one too hot, one too cold, and the third "just right." The interplay and balance of the biosphere, oceans, rock, air, and ice of Earth has kept conditions congenial for life for billions of years.

Carbon Dioxide and Temperature

Many of the links between Earth's atmospheric chemistry and its climate are complex and poorly understood. One connection, however, is crystal clear: For the last 160,000 years, at least, there is a consistent relationship between the amount of carbon dioxide in the air and the average temperature of the planet.

The importance of carbon dioxide in regulating Earth's temperature was confirmed by scientists working in eastern Antarctica. Drilling down into a glacier, they extracted a mile-long cylinder of ice from the hole. The glacier had formed as layer upon layer of snow accumulated year after year. Thus drilling into the ice was tantamount to drilling back through time.

Change in Temperature (Antarctic air temperature relative to today)

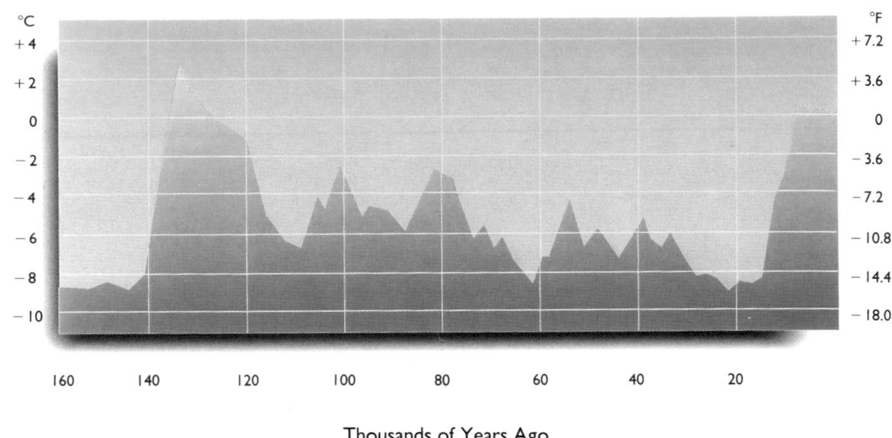

By extracting cylinders of ice from ancient Antarctic glaciers, scientists have constructed a 160,000-year record of air temperature and the atmospheric concentration of carbon dioxide. In the graphs at right and opposite, the two measurements are seen to rise and fall nearly in lock step.

The deepest sections of the core are composed of water that fell as snow 160,000 years ago. Scientists in Grenoble, France, fractured portions of the core and measured the composition of ancient air released from bubbles in the ice. Other instruments were used to measure the ratio of certain isotopes in the frozen water and thus to get an idea of the prevailing atmospheric temperature at the time that particular bit of water became locked in the glacier.

The result is a remarkable, unbroken record of both temperature and atmospheric levels of carbon dioxide; these two factors travel through time in lock step. Almost every time the chill of an ice age descended on the planet, carbon dioxide levels dropped. As the global temperature dropped 9°F (5°C), CO_2 levels dropped to 190 parts per million or so. Generally, as each ice age ended and the Earth basked in a warm interglacial, carbon dioxide levels rose to around 280 parts per million. Through the 160,000 years of that ice record, the level of carbon dioxide in the atmosphere fluctuated between 190 and 280 parts per million, but never rose much higher—not until the beginning of the Industrial Revolution in the last century.

There is indirect evidence that the link goes back much further than the glacial record. Carbon dioxide levels may have been much greater than the current concentration during the Carboniferous Period, which ended 285 million years ago. The period was named for a profusion of plant life whose buried remains produced a large fraction of the coal deposits that are being brought to the surface and burned today. Most land surfaces are thought to have been covered with lush swamps and bogs at the time. Coinciding with

Carbon Dioxide
(Parts per million,
by volume)

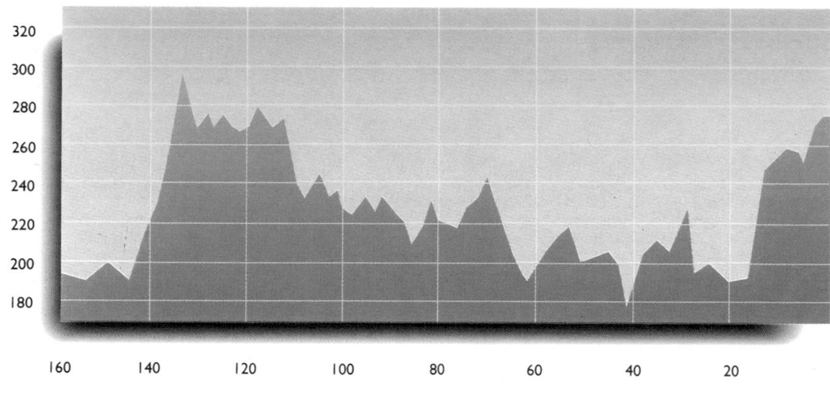

Thousands of Years Ago

the high CO_2 levels, global temperature was apparently also higher then, and there were no ice caps at the poles.

The causal relationship between carbon dioxide and temperature is still unclear; scientists do not know to what extent a drop in levels of carbon dioxide causes the cooling and to what extent cooling causes the change in carbon dioxide. But the two changes almost always go hand in hand. That is one reason scientists look with such concern at the recent rise in CO_2 produced by human activities.

Tampering with the Atmosphere

Starting perhaps as early as the mid-1700s—the time of the American Revolution, a time when Earth was still feeling the effects of the Little Ice Age—humanity began to fiddle with one of these two closely linked factors. The tampering probably started with the clearing and burning of vast tracts of forest in Europe and North America; for the first time, the atmosphere began to absorb large emanations of carbon dioxide from the activities of the growing human population. Then came steam engines and internal combustion engines and a cascade of technological developments resulting in new uses for heat—new uses for burning fuel and thus new sources of carbon dioxide. As forests fell, new and more efficient fuels were sought to replace firewood. From then on, most of the fuel would be extracted from the ground, in the form of coal, oil, and natural gas. Vast buried stores of carbon were uncovered, fed into the growing fires of the Industrial Revolution, and released into the atmosphere as carbon dioxide.

Starting with the Industrial Revolution, humans began to alter the landscape and to release enormous amounts of carbon dioxide into the air as coal, oil, and natural gas were burned. Above, a dragline shovel rips into a coal seam in a Montana strip mine.

As a result, before the end of the nineteenth century, the concentration of carbon dioxide in the atmosphere was already approaching 300 parts per million and steadily rising. Soon it had risen beyond the highest point it had reached in at least 160,000 years.

Although no one at the time had any way to measure the change, a Swedish chemist named Svante Arrhenius theorized that this change was occurring. Arrhenius was the first scientist to see the significance of the greenhouse theories of Fourier and Tyndall in light of events of the late nineteenth century. As he looked around at the growing forests of chimneys and smokestacks, the steam engines, furnaces, foundries, and ovens—all stoked with coal, charcoal, and wood—he calculated that millions of tons of carbon dioxide were being released. He did not know of the historical link between carbon dioxide and warmth that would later be discerned deep in glacial ice. He did know that in theory all that carbon dioxide, by causing a "change in the transparency of the atmosphere," as he put it, could very likely heat things up. A doubling of carbon dioxide, he found, might raise the average temperature of the planet 9°F (5°C).

In an essay in the April 1896 issue of the *London, Edinburgh, and Dublin Philosophical Magazine*, Arrhenius spelled out his theory—and in a nine-word sentence he aptly summed up the hidden consequence of the rapid expansion of human industry that was unbalancing the atmosphere. Arrhenius, who would seven years later win one of the first Nobel Prizes in chemistry (for his theory of ionization), wrote: "We are evaporating our coal mines into the air."

THE HAND OF MAN

"I'm truly sorry Man's dominion
Has broken Nature's social union. . . .
The best laid schemes o' mice an' men
Gang aft a-gley [often go awry]."

—Robert Burns, Scottish poet (1759–1796), "To a Mouse"

Almost a century has passed since Arrhenius first posited that human actions were changing the nature of the world. The dizzying pace of change in these hundred years only becomes evident when you look at how much some places have been modified in only a few generations. Just consider this account of one American wilderness, as described in the 1875 book *Fishing in American Waters*, by Senio C. Scott: "There is not within any settled portion of the United States another piece of territory where the trout streams are so numerous and productive. . . . It is scarcely possible to travel a mile in any direction without crossing a trout stream." The place? Long Island, New York.

"From Coney Island to Southampton," Scott wrote, there was one clear, trout-laden stream after another. Today the only trout on Coney Island are in the smoked-fish sections of the Russian delicatessens on Mermaid Avenue.

It is now hard to find any place on Earth where the human impact is not evident. Half a century ago, if you stood on a hilltop on a clear day just about anywhere east of the Rocky Mountains, you could have seen things seventy miles away. Now, average visibility—even far from cities—is about fifteen miles. Particulates from power plants and automobiles have created a permanent haze.

Even high above the stark, frozen ice fields of Antarctica, satellites and research aircraft now routinely detect the seasonal formation of a gaping hole in the thin veneer of ozone that has shielded terrestrial organisms from harmful ultraviolet radiation ever since life first spread onto dry land half a billion years ago. And scientists have confirmed that the degradation of the ozone layer is being caused by man-made chlorofluorocarbons, the same CFCs that are also powerful greenhouse gases.

Road-building projects have lured hundreds of thousands of small farmers and ranchers into the lush Amazon rain forest (opposite). Wherever roads are built, trees fall.

Rain forests are vanishing at the rate of one and a half football fields each second. On the preceding pages, an Indonesian farmer cuts brush in a smoldering landscape.

A Staggering Rate of Change

As a journalist, I have been fortunate to travel to some of the more remote corners of the planet, and nowhere have I found a place unaffected by human actions. The scars are perhaps freshest in the Amazon. I remember bouncing

Bulldozers scrape away soil along the Trans-Amazon Highway of Brazil (left). Massive carbon dioxide emissions from deforestation put Brazil near the top of the list of largest greenhouse gas polluters.

Resettlement programs in Indonesia have lured people from crowded islands, such as Java, to less populated ones, like Sumatra (opposite). One result: rain forests are being slashed and burned at a frenzied pace.

around one evening in the back of a small pickup truck as it sped along one of the muddy roads that have been cut through that vast, forested river basin. Just twenty years ago, another passenger told me, the rich rain forest along that road was pristine; in some places leafy branches formed a vaulted gallery over what was then a rough dirt trail. Now that a flood of landless poor and cattle ranchers had swept through, the forest was cut back so far that it was only a faint green smudge on the horizon. Here and there stood a single giant tree trunk topped by dead branches or thin foliage—a pathetic vestige of the vanished forest.

Worldwide, rain forests are disappearing at the rate of one and a half football fields a second. Just a few centuries ago, Earth's equator was girdled by a green belt of 15 million square miles of rain forest, an area five times that of the contiguous United States. Now three Americas' worth of forest are gone, with just 6.2 million square miles left. In Brazil in the late 1980s, the annual emission of carbon dioxide from the burning of forests equalled the amount of this gas spewing from the industries of Poland and West Germany combined. Because of the burning, Brazil was fourth on the list of greenhouse polluters, behind the United States, the Soviet Union, and China. Without the burning, Brazil would not even be in the top twenty polluters. Alberto Setzer, a Brazilian space scientist who monitored the fires using satellite photographs —sometimes counting more than eight thousand fires in a single day—

The destruction of rain forests not only adds significantly to the greenhouse problem, but also threatens countless species of plants and animals with extinction. Above, an iguana clings to a tree in the Ecuadorian Amazon. Right, poison-arrow frogs nestle in a toadstool in Costa Rica's Corcovado National Park. Opposite, clockwise from bottom: a pair of blue-winged macaws perch in the branches of Brazil's coastal forest; a passion flower blooms in Costa Rica; and an exotic butterfly (family Riodinidae) sits on a leaf in a Sumatran rain forest.

calculated that emissions from the annual burning season in the Amazon equalled those of a large volcano. But, as Setzer put it, "This is a volcano that erupts every year, not just once in a lifetime."

Back in 1980, I was working on a boat that sailed up the Red Sea. We passed a maze of oil rigs and pipelines, with natural gas burning off in towering plumes of flame and black smoke. An endless convoy of ships headed north for Europe, some riding low—their holds filled with oil—and others riding as high as a ten-story building, stacked with layer upon layer of Japanese automobiles. Once at their destination, these two imports—oil and cars—would meet, and the result would be more carbon dioxide and smog.

Off the coast of Ethiopia, we passed a string of uninhabited islands that were about as bleak and sterile as any terrain on Earth—blasted volcanic heaps with hardly a shrub growing in the gray and black soil. The islands had risen along a great submarine rift where Africa is slowly tearing away from the Eurasian continent. Nothing looked odd as we anchored off Zuqar Island. The coral below glimmered, and a school of manta rays, each as long and broad as a king-size bed, soared through the clear water. But as we came ashore and explored the beach, we were stunned by what we saw.

Hundreds of light bulbs had somehow bounced their way above the tide line without shattering. Bulbs littered the shoreline as far as we could see.

In the Red Sea, oil production and pollution from passing ships mar the landscape and threaten fragile coral reefs. At left, an oil rig in the Gulf of Suez burns waste gases. The stark volcanic shores of Red Sea islands such as Zubayr (opposite, top) give way to a biological wonderland beneath the waves. A box fish pokes from a coral crevice (opposite, bottom left), and fish swarm around a Red Sea reef (opposite, bottom right).

Below, smog shrouds the skyline of Los Angeles, a city that has grown up around its network of freeways.

There were long fluorescent tubes and high-wattage spotlights and ordinary screw-in incandescent bulbs. We figured that most were burned-out bulbs that had been tossed from passing ships. Hundreds of miles from the nearest town, we were surrounded by the light bulb, one of the crowning symbols of technological progress.

Closer to civilization, things were far worse. I lived for a time in Los Angeles, where I reported on the environmental problems plaguing one of America's fastest-growing cities. From the hills above Hollywood Boulevard, where I lived, I could look out across the sea of smog that engulfs that city most days. It has been just four years since I left Los Angeles, but in that time, the average speed on the freeways has already fallen from 38 to 35 miles per hour. (It is projected that, by 2010, at the current rate of growth, traffic could creep along at 19 miles per hour.)

While in Los Angeles, I wrote about some strange events, including a discovery made one day by a construction crew. They were excavating a site, preparing to pour the foundation for a parking garage. Suddenly gasoline began bubbling up from the earth. They had not struck some underground

Automobiles and trucks consume more than 60 percent of the 17 million barrels of oil burned in the United States each day, adding hundreds of thousands of tons of carbon dioxide to the global greenhouse. At right, rush-hour traffic creeps across the Bay Bridge connecting the California cities of Oakland and San Francisco.

pipeline, but instead had simply dug down to the water table. Later, authorities found that the gas had been leaking for years from a storage tank at a service station several hundred yards away. Floating on one of the only big pockets of groundwater in a desiccated city aching for water was a spreading, subterranean lake of gasoline.

We live in a time when the natural landscape, in places, is scarred almost beyond recognition. Such scars are a disturbing presence, but now it is the less obvious impact of man—Arrhenius's "change in the transparency" of the air—that has scientists most concerned. In the century that has passed since he wrote about all that "evaporating" coal, the sphere of influence of the human species has undergone one of the most explosive expansions ever seen in the long history of life's adventure on Earth.

"A Great, Roaring, Wasteful Furnace"

In the year A.D. 1, there were about 250 million human beings—the current population of the United States, spread across a planet. It took 1,650 years for that number to double. Between 1650 and 1930, the human population

An aerial view shows the expanse of an open-pit coal mine in Montana. Between 1870 and 1970, the burning of fossil fuels such as coal added 400 billion tons of carbon dioxide to the air. From 1970 to 1989, just 19 years, another 400 billion tons was added.

Just a little over a century ago, oil was a novelty. The photograph directly above, taken circa 1866, shows the world's first oil well, near Titusville, Pennsylvania. The well was dubbed the Drake Well, after its owner, E. L. Drake, seen wearing the stovepipe hat. In a picture taken in 1919 (above right), a horse-drawn coal train leaves a mine in Scranton, Pennsylvania.

rose fourfold, to 2 billion. By the end of the twentieth century, there will be more than 6 billion people on Earth—triple the population 70 years earlier.

In this century, it has not just been rising numbers that have increased human dominion of the globe, but an increase in humankind's control, through technology, of all the resources of the planet. In this short span, there have been extraordinary advances—particularly in the technology of combustion.

In his 1954 essay "Man the Firemaker," Loren Eiseley correlates human progress with the use of ever-more-potent fuels. First came firewood, which enabled humans to cook meats and thus increase food's nutritive value. Then came charcoal. The Iron Age would have been meaningless without the hot charcoal fires over which metals become malleable. Mastery of glass, ceramics, and steel was a function of rising temperatures in kilns, forges, and furnaces.

As Eiseley put it, "Man's long adventure with knowledge has, to a very marked degree, been a climb up the heat ladder. . . . Today the flames grow hotter in the furnaces. . . . The creature that crept furred through the glitter of blue glacial nights lives surrounded by the hiss of steam, the roar of engines, and the bubbling of vats. . . . And he is himself a flame—a great, roaring, wasteful furnace devouring irreplaceable substances of the earth."

In the words of the Russian geochemist Vladimir Vernadsky, it is only in the twentieth century that humanity "for the first time becomes a large-scale geological force."

Fans crowd the stands at a soccer match (above). The Earth's human population is expected to reach 6 billion by the end of this decade, and 10 billion by the middle of the coming century. As all of those people clamor for space, fuel, and goods, the planet and atmosphere will have to endure more assaults.

A large part of that force has come from the burning of fuels. Since 1900, annual consumption of energy has increased more than 10-fold. Between 1870 and 1970, the burning of fossil fuels added 400 billion tons of carbon dioxide to the air. From 1970 to 1989—just 19 years—another 400 billion tons were added. These numbers give some indication of the stunning acceleration of the human impact. As we approach the year 2000, another 60 million tons of carbon dioxide are spewn into the air every day.

Some comes from the exhaust pipes of the 500 million cars on the planet. (By 2030, the automobile population is expected to double—1 billion cars!) Some carbon dioxide comes from the furnaces of steel mills, cement plants, and power plants. Some comes from fires set to clear brush and create cattle pasture. Part of that human production of carbon dioxide comes from cooking fires—more than two billion people cook their daily meals over firewood.

Carbon Dioxide (Parts per million, by volume)

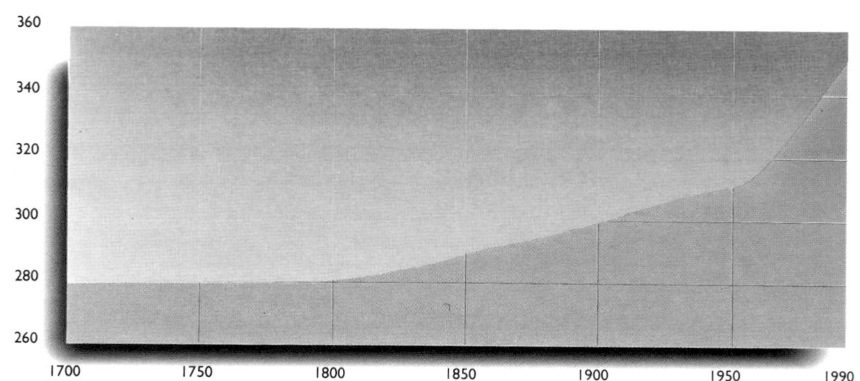

As the Industrial Revolution got underway in the mid 1700s, human beings began to alter the atmosphere. The concentration of carbon dioxide rises particularly sharply from 1950 onward. This reflects the explosion of fossil fuel use in the last 40 years.

A Turkana nomad in Africa prepares corn porridge over an open fire (right). More than 2 billion people rely on wood to cook their daily meals. Open fires are very inefficient and contribute significantly to carbon dioxide emissions.

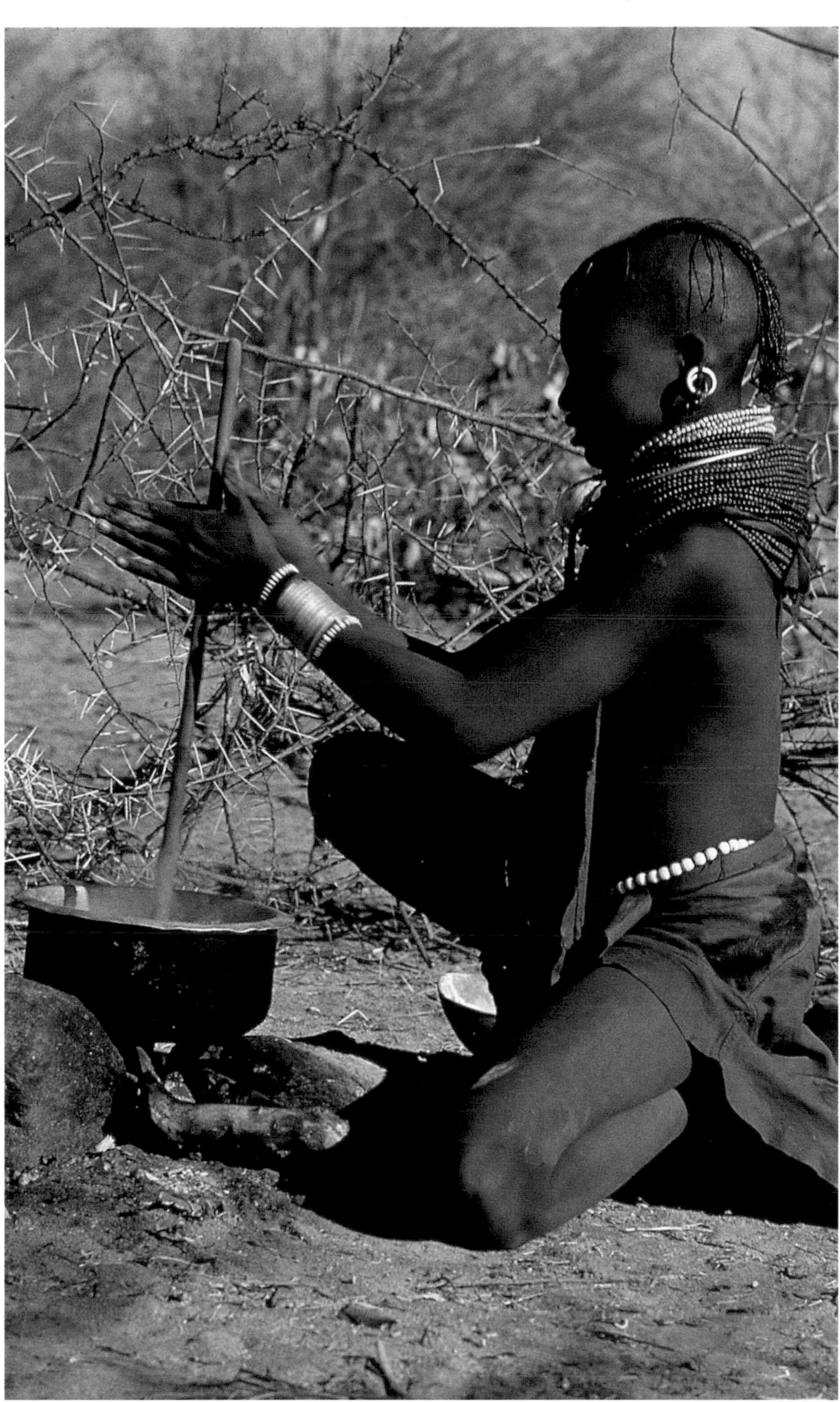

This image, a composite of satellite data, shows Earth at night. Most of the bright spots over central Africa are not electric lights, but fires set by people. It is estimated that 2 to 5 percent of the planet's land surface burns each year, adding further to the greenhouse effect.

All in all, then, as humans have burned things, they have added a lot of carbon dioxide to the air—some 850 billion tons in a little over a century. When you consider that the entire atmosphere weighs some five million billion tons, this number seems less disturbing. It quickly grows in significance, however, when you recall that a little carbon dioxide accounts for a significant portion of the air's warming effect.

A Nation of Pharaohs

In 1981 the biologist and philosopher René Dubos railed at the excesses of our carbon-fueled civilization, in which so much growth has been built on "a series of great technological achievements made possible by the lavish use of cheap fossil fuels."

Today, every person with an automobile has the power of a king, he wrote in *Celebrations of Life*: "Personal control of a 350-horsepower automobile is equivalent in energy terms to the power of an Egyptian pharaoh with

350 horses or 3,500 slaves at his command." Even accounting for those who drive more conservative vehicles, with less than 150 horsepower, that still makes countries such as the United States nations of pharaohs. Not surprisingly, the rest of the world now wants to catch up.

Awakening Scientific Interest

After Arrhenius, there was some continuing interest in the idea that humans were modifying the atmosphere in ways that could warm the planet. In the 1930s, during a period of unusually hot summers in Europe, George Callendar, a British coal engineer, compiled several decades' worth of temperature readings taken at dozens of weather stations and by sea captains who brought up buckets of water and measured its warmth. He averaged the readings and published his results in 1938. His graphs showed a steady warming trend, which he ascribed to rising carbon dioxide levels. Callendar optimistically asserted that the addition of carbon dioxide to the air would result in a warmer, more comfortable world. (Arrhenius had earlier drawn this same conclusion, once writing that we "may hope to enjoy ages with more equable and better climates.") Callendar's theories were forgotten as, shortly thereafter, the northern hemisphere experienced a prolonged cold period. Once again, the vagaries of year-to-year variations in weather had forestalled efforts to understand the deeper workings of climate.

Another reason that the rise in carbon dioxide levels was largely ignored was that basic laws of gas exchange led people to believe that the oceans would act as a vast "sink," or repository, for this gas. Scientists calculated that nearly 98 percent of any excess carbon dioxide in the atmosphere would be absorbed by the oceans, and the carbon would soon find its way to the sea floor—where it would be locked safely away in sediments for millions of years to come.

This view prevailed until 1957, when Roger Revelle announced his analysis of the question. Revelle, then director of the Scripps Institution of Oceanography in La Jolla, California, made some calculations and found that the chemistry of seawater limited the amount of carbon dioxide that would dissolve in it. Although much of the gas would end up in the seas, Revelle calculated that as more and more carbon dioxide was produced by human actions, a significant amount would remain in the air—more than enough to influence the workings of the atmosphere. As Revelle and his coauthor, Hans Suess, put it: "Thus human beings are now carrying out a large-scale geophysical experiment of a kind that could not have happened in the past nor be reproduced in the future."

"Large-scale" was perhaps the understatement of the century. The test tube in which this experiment was taking place was the entire planet.

Around that time, Revelle was involved in planning the most ambitious coordinated scientific examination in history of the workings of the globe, the International Geophysical Year. Actually running over eighteen months, the project would employ hundreds of geologists, chemists, climatologists, and other specialists from more than seventy nations. They would be dispatched to the four corners of the Earth, where they would measure everything measurable. In a way, they were about to conduct Earth's first checkup.

Hard Evidence: The Keeling Curve

One of the scientists was a young chemist named Charles David Keeling, who had recently spent two years running around California collecting flasks full of air. Keeling had designed and built an especially sensitive instrument for measuring concentrations of carbon dioxide; his was the first device that could measure differences on the order of one part per million. Keeling was finding that his gas samples from around the state contained, on average, 315 parts per million of CO_2—a figure about 13 percent higher than the few measurements that were available from before the Industrial Revolution.

As part of the International Geophysical Year, Keeling built two instru-

At this weather station near the summit of the Mauna Loa volcano in Hawaii, scientists have been measuring the levels of carbon dioxide in the air since 1958. The concentration has inexorably risen.

ments, called manometers, that could take continual readings of atmospheric CO_2. One was hauled up the slope of Mauna Loa, a massive dormant volcano that crowns the big island of Hawaii. There, 11,050 feet above sea level—far from the distorting influence of cities and forests—the manometer began in March 1958 to take readings of the carbon dioxide level of the atmosphere.

In its first year, the instrument produced a remarkable record of the breathing of the biosphere. A graph of the readings looked like a series of camel's humps. Carbon dioxide was high in the winter, then dropped in the summer, then rose again in the fall and the following winter. Keeling's meter was reading the annual rhythm of photosynthesis in the Northern Hemisphere. In the warmth of the spring and summer, the temperate forests burst into photosynthetic activity. In the process, the trees removed millions of tons of carbon from the air and put it into new growth—roots and stems and blossoms and berries and pinecones. With winter came torpor; the plants of the Northern Hemisphere released much of that carbon as leaves fell and disintegrated, fruit rotted, and trees consumed some of their stores of energy. The following year, and the year after that, the pattern was repeated.

Keeling became fascinated with carbon dioxide's cycles, and he kept his instruments running long after the big research project was over. For three decades his first device, and many others like it that were later deployed to widely dispersed weather stations, have been churning out readings of the atmosphere's carbon dioxide trace. The result, as all the data have been collated and assembled in a graph, is a remarkable, sinuous curve, dubbed the Keeling Curve, that may prove to be a central symbol for the Anthrocene Age, our legacy on Earth.

Each year's curve has the same camel's-hump shape; but as the numbers for successive years are strung together, the line of the graph moves steadily up the page—a snake climbing a long hill. The levels of carbon dioxide have been just a little bit higher each year than they were the year before.

Slow, Steady—and Suspicious?

The process is hardly discernible—like a watch that gains a few seconds a month—but the end result is significant. In 1958, the atmosphere had 315 parts per million of carbon dioxide; by 1990 the level was 355 parts per million of this heat-trapping gas. That increase means that since the 1880s, the amount of carbon dioxide in the atmosphere had risen 22 percent. What is more, each year the curve is climbing at a faster rate. Whereas in the 1960s, Keeling found that carbon dioxide levels were rising about 0.7 parts per million a year, by the 1980s the annual rise had more than doubled, to 1.5 parts

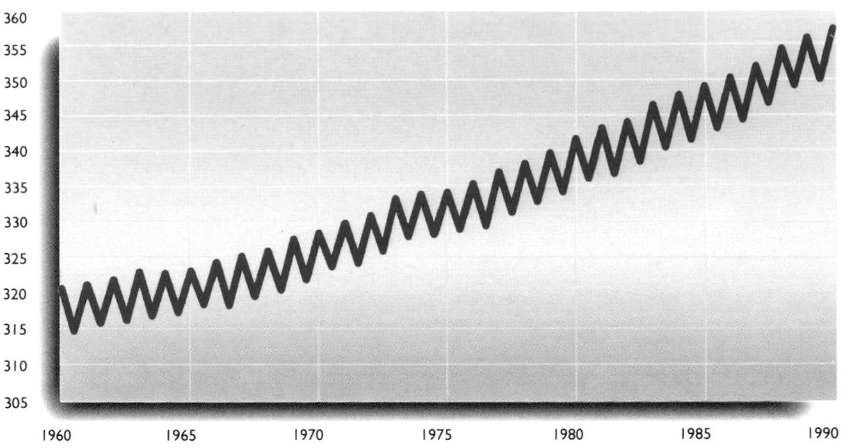

Carbon Dioxide (Parts per million, by volume)

Measurements of atmospheric carbon dioxide show a steady rise in the level of this heat-trapping gas from 1960 to 1990. This curve was named the Keeling curve, after the man who initiated the measurements. The concentration of carbon dioxide has risen 12 percent in those 30 years and has increased a total of 25 percent since the beginning of the Industrial Revolution. The annual downward flutter in the curve begins as vegetation in the Northern Hemisphere grows in summer, drawing carbon dioxide from the air through photosynthesis.

per million a year. This rise reflects the enormous acceleration of global industry and the acceleration of population growth, both of which were adding tremendously to the amount of burning.

Keeling, like Callendar before him, happened to begin his research at a time when global temperatures were tending to dip. Thus, his results were widely discussed among climatologists, but rarely cited as a cause for concern. Indeed, the 1970s saw prolonged periods of wintry weather. Brazil's coffee crop failed twice because of frosts. The United States saw freak blizzards. Snow fell in London in June. The chilly weather had politicians and the press buzzing about an imminent ice age.

In the late 1970s, though, the global average temperature resumed the climb it had begun back in the 1930s. Year after year a new record was set, as readings from around the world were collected and corrected—to remove distortions caused by phenomena such as the so-called "heat-island" effect of cities (which hold heat in their masses of stone and asphalt and thus tend to make local average temperature readings higher than they would otherwise be). In 1988, researchers at NASA's Goddard Institute for Space Studies and at the University of East Anglia in England separately reported that when all such distortions were removed, the global mean temperature had risen almost 1°F (.5°C) in 100 years. The rise was uneven, but compared to anything in the past 10,000 years, it was happening at breakneck speed. There was no way to prove a causal relationship between the rise in carbon dioxide and the rise in temperature. Indeed, the warming was within the range of normal fluctuations of temperature and could have been caused by any number of factors, including variations in the sun's output, changes in volcanic activity, or random fluctuations in climate. Nonetheless, the good match between the carbon dioxide curve and the temperature curve began to generate more and more interest in the scientific community and the world at large. The distinctive climbing snake of the Keeling Curve began to show up not just in scientific journals, but also in the pages of the *New York Times*, *Newsweek*, and *National Geographic* magazine.

CO_2: Not the Only Villain

Moreover, concern about an intensifying greenhouse effect was heightened when it became apparent that human activities were increasing the levels of other trace gases that have much more potent heat-trapping properties than CO_2. Levels of methane, for example, which is twenty to thirty times more efficient at trapping heat, have risen sharply in the atmosphere in the past 150 years. Although the methane concentration is still below 2 parts per million, that is more than double the highest level seen for the past 160,000 years. Methane—also called natural gas or swamp gas—is generated naturally by bacterial decomposition in the absence of oxygen, in such places as landfills, bogs, rice paddies, and the guts of cattle and termites (where it is produced when anaerobic bacteria digest cellulose). As humans have cut down forests, termites and bacteria have flourished in the detritus left behind. Moreover, the human population explosion has been accompanied by a similar increase in livestock: there are some 1.3 billion head of cattle on Earth today. Each cow belches methane, on average, about twice a minute.

Methane also escapes during coal mining and oil drilling; it is the gas that burns in the flares above oil rigs. Some methane is also emitted by the incom-

Proportional Contribution of Increased Greenhouse Gases to Global Warming

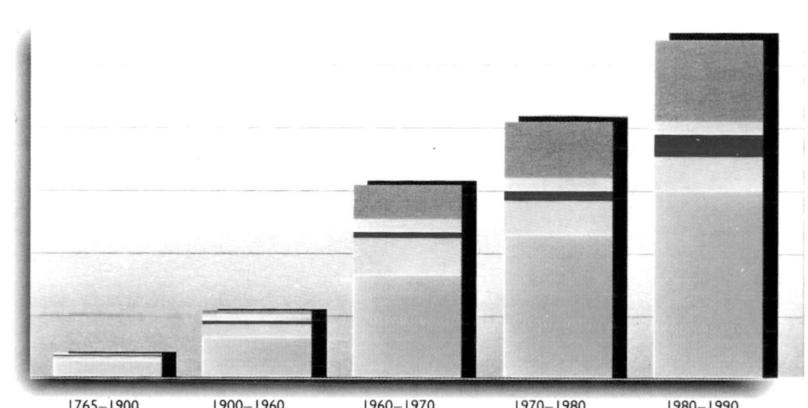

Chlorofluorocarbons (CFCs)
Stratospheric Water Vapor
Nitrous Oxide
Methane
Carbon Dioxide

Carbon dioxide is just one of the gases adding to the greenhouse effect. This graph shows the relative contribution of several key gases: carbon dioxide, methane, nitrous oxide, stratospheric water vapor, and chlorofluorocarbons (CFCs). By the middle of the next century, the greenhouse impact of the other gases combined may equal that of carbon dioxide.

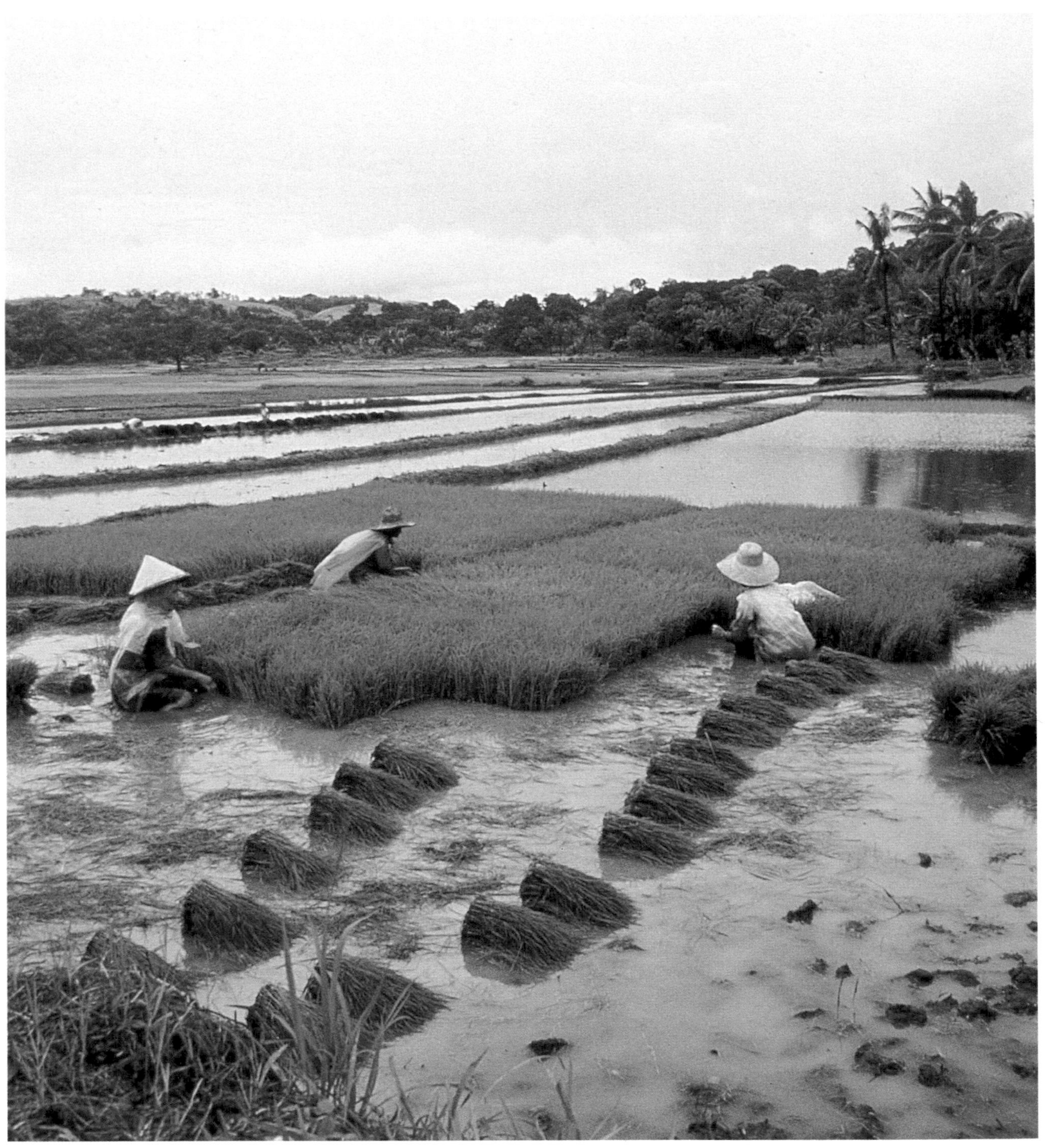

A major source of methane is anaerobic bacteria, which thrive in rice paddies such as these (opposite), some 100 miles north of Manila in the Philippines. Methane-producing bacteria also live in the guts of cattle (right). As human populations have risen, so, too, has the number of cattle, which today is about 1.3 billion.

plete combustion of fossil fuels. All told, about half of the methane in the atmosphere today is thought to be the result of human activity.

Some scientists theorize that part of the methane increase is the result of the slight global warming that has already taken place: Methane may be rising from warming soil and thawing permafrost in the vast stretches of Arctic tundra, and perhaps from the floor of the sea, where huge quantities of methane are locked up in cold sediments. If rising temperatures allow a large portion of this hidden reserve of methane to enter the atmosphere, then the greenhouse effect will accelerate.

Another gas on the rise is nitrous oxide, the same compound that some dentists use to dull pain. In nature, nitrous oxide is emitted by soil microbes. When nitrate fertilizers are added to farmland, they accelerate the production of the gas. Sewage disposal and deforestation may also lead to increased emissions. Altogether, there has been a rise of 8 percent in levels of nitrous oxide since the turn of the century.

Nitrous oxide, methane, and carbon dioxide have all been around, to some extent, for millions of years. But there is a fourth important class of greenhouse gases that never existed in nature before this century. These gases are chlorofluorocarbons (CFCs), and were invented in 1930 by Thomas Midgley, Jr., a remarkably creative chemist working for General Motors. Midgley had already made a name for himself by concocting the tetraethyl-lead gasoline additive that took the knock out of auto engines. He was asked by GM's Frigidaire division to come up with an alternative to ammonia and sulfur

Nitrate fertilizers are a source of nitrous oxide, one of the fast-increasing greenhouse gases. Above, ammonium nitrate is applied to a wheat field in McCook, Nebraska.

dioxide, which were then the standard chemicals circulating in the coils of refrigeration units—but were dangerous and toxic. Midgley's solution was to add some chlorine and fluorine atoms to a carbon chain: the result was an inert, nontoxic class of compounds called chlorofluorocarbons—also known by the trade name Freon. These CFCs were hailed as one of the great triumphs of the chemical age—a purely synthetic substance that lasted for centuries and had no known adverse effects. Midgley won important awards for both his lead additive and for CFCs, which quickly found myriad uses in products ranging from plastic foams to underarm sprays.

Ironically, these two substances now occupy the highest levels of environmental hazard lists. Today, lead levels in human tissue and Arctic snows are hundreds of times what they were two centuries ago. And CFCs are not

Discarded refrigerators and air conditioners contain chlorofluorocarbons (CFCs) that will eventually leak into the atmosphere, adding greatly to the greenhouse effect and also eroding Earth's protective ozone layer.

only a potent, long-lived greenhouse gas, but they are also the compounds that have wafted up to the stratosphere and attacked the planet's protective shield of ozone. Although CFCs exist in the atmosphere at minute levels—measured in parts per *trillion*—each molecule has more than 12,000 times the heat-trapping potential of a molecule of carbon dioxide. Chlorofluorocarbons have pervaded industry. In sixty years, 16 million tons of them have made their way into the air. Even though the United States banned some uses in the 1970s, and many countries have agreed to eliminate production by the turn of the century, the amount in the air is still growing at some 5 percent each year. And each molecule may remain in the atmosphere for many decades. Midgley's chemical creation is certainly a durable one.

The sudden accumulation of carbon dioxide, methane, nitrous oxide, and CFCs is not visible. You cannot smell it because none of the gases has an odor (the natural gas in a kitchen stove has an odorant added to it to warn of a leak). But this flurry of atmospheric change is bound to disrupt the balance of energy flowing to and from Earth, and, as a consequence, the balance of the biological world as well.

Graphs of the sharply ascending amounts of the rarer greenhouse gases, taken together with the rising snake of the Keeling Curve and the rise in the century-long graph of global temperature, are reminiscent of the charts at the bedside of a patient in intensive care. They indicate that the planet's vital signs are wavering. The challenge is that by the time the symptoms are understood, the patient may be gravely ill indeed.

THE CLOUDY CRYSTAL BALL

"How little, mark! that portion of the ball,
Where faint at best, the beams of Science fall."

—Alexander Pope, English poet (1688–1744), "The Dunciad"

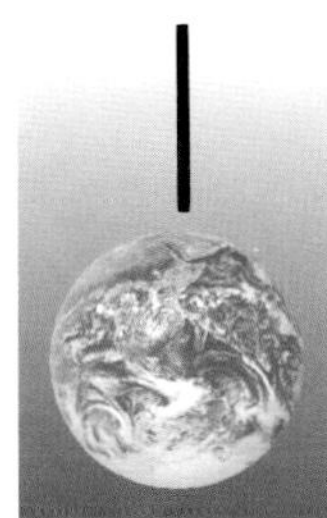

It is July 2029, and an atmospheric catastrophe is in the making. The amount of carbon dioxide in the air has nearly doubled from pre-industrial levels. Earth is running a raging fever. Over the American Grain Belt and central China, the heat of the sun withers wheat and corn fields. Over the southern oceans around Antarctica, air temperatures are far above the typical readings seen a hundred years earlier. As a result, icebergs the size of Rhode Island are calving from the edges of glaciers at a record rate. As the seasons and years roll forward, the atmospheric concentrations of carbon dioxide and other greenhouse gases continue to rise. Heat waves and occasional cool snaps swirl around the planet, but by 2050 the colder pockets are a rarity, and the entire globe is baking in conditions many degrees warmer than were typical just a century before.

Fortunately this is not yet the real world, but merely a projection, built of a matrix of mathematical equations, each representing conditions in a small portion of the atmosphere. The fate of this world is unfolding on a video display terminal in an air-conditioned computer room two floors above Tom's Restaurant, a luncheonette at the corner of Broadway and 112th Street in Manhattan. That setting is the unlikely home of the Goddard Institute for Space Studies, a research center run by NASA that has focused for more than a decade on the question of climate change. As time rolls forward in the computer model, swirls of red, orange, yellow, white, and blue are painted on the screen, where Earth's continents are visible in outline. As summer follows summer, the heartland of America pulses cherry red—a color that represents temperatures 9°F (5°C) above today's norm. The orange and yellow indicate less intense warming, and the few blue patches show rare regions that remain relatively cool.

Such global climate models, also called general-circulation models, were first developed both to help forecast the weather on a daily basis and to reveal the workings of the atmosphere of other planets (hence the NASA connection). The picture the models project of Earth beneath the intensified greenhouse is not a pretty one.

In the Arctic, global warming is expected to raise temperatures dramatically, causing increased melting of glaciers and calving of icebergs. Opposite, meltwater cascades from the Holgate Glacier in Kenai Fjords National Park, Alaska.

On the preceding pages, crabeater seals bask on sea ice in Antarctica. Some scientists theorize that warming seas may hasten the breakup of vast glaciers there. Others calculate that additional precipitation will fall on Antarctica in a warmer world, thickening glaciers.

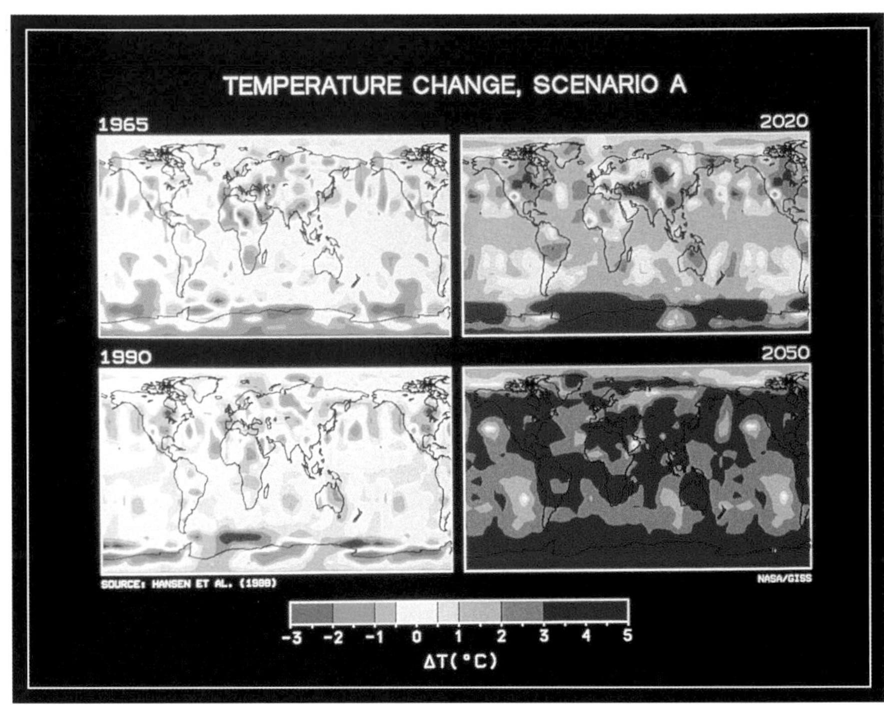

When rising carbon dioxide levels in the atmosphere are fed into computer models of global climate, the simulated world begins to warm. In this model, run by NASA scientists, very little warming is evident from 1965 to 1990—matching what has happened in the real world. But from 1990 on, temperatures begin to rise almost everywhere. (The hotter temperatures are represented by warmer colors, as indicated on the scale at the bottom; temperatures are given in degrees Centigrade.) The projection shows that by 2050, Earth will, on average, become a much warmer place.

If our planet were a smooth, motionless sphere, the task of simulating its climate would be simple. You would need only equations for the amount of sunlight hitting the surface and the infrared radiation escaping into space. There would be no regional variations. There would be no wind or clouds or ocean currents or polar ice caps to stir things up in unpredictable ways. Earth, however, is a variegated, complicated system with a mottled surface of water, earth, ice, and green plant life—each of which absorbs sunlight and radiates heat in very different ways. As a result, modeling future climates is a daunting challenge. Even so, computer programmers and atmospheric scientists have reached a point where their mathematical simulations do a remarkably good job of representing the real world.

How the Computer Simulations Work

The global climate models at research centers such as the Goddard Institute all work in much the same way. The atmosphere is crudely represented by splitting it into layers, like those of an onion, and then "dicing" each section. A typical model divides the atmosphere vertically into about a dozen layers, each a mile or so thick, and horizontally into boxes that are roughly the size of

Colorado. The conditions in each box are boiled down to a few basic equations that represent the flow of heat, moisture, sunlight, wind, and the like. Each box in the atmospheric grid interacts with adjacent boxes—allowing, say, a parcel of hot air to rise through the layers of the atmosphere. The computer program accounts for the passage of the seasons, with the tilt of the Earth's axis as it revolves around the sun bringing less and then more sunlight to boxes in each hemisphere.

To get the program going, scientists feed into the computer a set of initial conditions—such as the conditions that prevailed globally in 1958, the first year that global carbon dioxide levels were measured. Then the scientists sit back and let the equations churn away. Program operators can tweak a particular condition—they can, say, add carbon dioxide to the air—by changing the appropriate factors in the set of equations. After all, carbon dioxide behaves in a predictable way, consistent with the basic laws of physics. It does not directly affect moisture or sunlight, but it does trap heat. So a rise in carbon dioxide is represented by changing the equations that indicate how much heat can be trapped in the air.

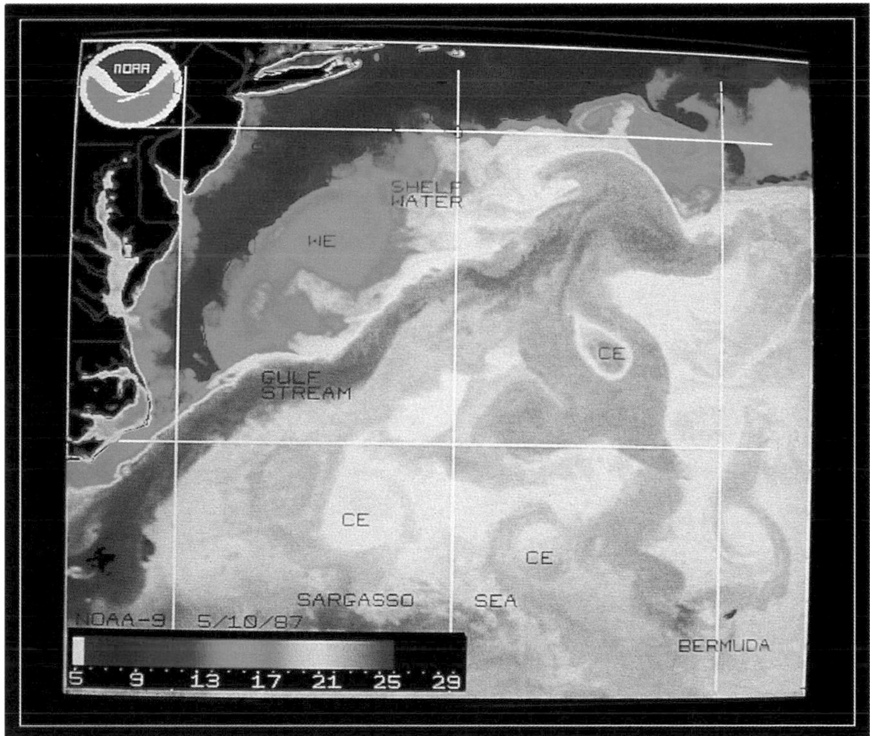

The oceans store and transport enormous amounts of the sun's heat, adding to uncertainties about how global warming will manifest itself. This image, created from satellite measurements of water temperature, shows how the Gulf Stream moves warm water north along the Atlantic coast (note Long Island near top left). The scale at the bottom shows water temperature in degrees Centigrade.

Changes in ocean temperatures and currents can strongly affect the web of life, causing blooms or die-offs of plankton and other sensitive marine species. Adelie penguins (above) feed on shrimp-like organisms that in turn feed on plankton, and thus may indirectly feel the peril of global warming.

Originally, all of the projections of a hotter future were made by taking an operating climate model and immediately doubling the amount of carbon dioxide in the atmosphere. More recently, modelers at Goddard and other centers have been able to simulate a gradual addition of carbon dioxide in a way that more closely resembles the buildup of the gas in the real world. This calculation is very complex. At research centers that are equipped with the latest supercomputers—such as the National Center for Atmospheric Research in Boulder, Colorado—the fate of the world is determined in a matter of hours or a few days. At Goddard, which uses slower conventional computers, some "runs" have taken years to complete.

Many uncertainties still limit the confidence that can be placed in the model results. After all, the models remain very rough projections of reality. Each Colorado-size grid box can only be all cloudy or all sunny, for example—when it is obvious to anyone that conditions in Denver at one moment are likely to be very different from those in Durango.

Along with these limits of resolution, there are great uncertainties concerning the role of oceans and clouds in maintaining the current climate and affecting possible changes down the line. Two-thirds of the planet's surface is covered by oceans, which are a vast reservoir for heat and also act to transport that heat from the hot tropics to higher, cooler latitudes. The warm Gulf Stream, for example, is like a fast-flowing river—with a hundred times the volume of the Amazon—that swings north past Florida and then out across the Atlantic, carrying solar energy absorbed near the equator as far north as the British Isles and northern Europe. As a result, those areas are much warmer in winter than they would otherwise be; London, for instance, lies at the same latitude as the blustery northern tip of Newfoundland. And the Scilly Isles, off the southwest coast of England, have palm trees.

No one has a good idea of how the oceans may accelerate or slow the warming trend, or how ocean currents may affect the regional impact of global warming. We do know, however, that changes in the oceans can cause widespread changes in climate. Every few years, El Niño, a warming of the eastern Pacific near the Equator, can strongly alter rainfall and temperature patterns across the United States and as far east as Africa and the Indian Ocean. Recently, there have been attempts to link computer models of ocean currents to the global climate models, but the work is at an early stage.

Clouds' Influence on Climate

Clouds are one of the trickiest unknowns in the formula for global warming. They are a prominent feature of the planet, covering about 60 percent of

Clouds strongly influence climate, both by trapping some of the planet's heat and by reflecting some of the sun's rays from their white tops—they both warm and cool the planet. Opposite, clockwise from top: A view of the Hawaiian islands from space shows the sun reflecting from cloud tops; puffy cumulus clouds tend to reflect more energy than they trap, exerting a net cooling effect on the planet; high, wispy cirrus clouds trap more energy than they reflect, and thus tend to warm the Earth.

Earth's surface at any given time. But their influence on climate is hard to add up. These floating blankets of water droplets or ice crystals simultaneously reflect sunlight back out to space and trap escaping heat—they both cool and warm the planet. Moreover, different types of clouds have different effects. High, wispy "mare's tails" tend to have a net warming effect—they trap more energy than they reflect. Low, thick cumulus clouds have a net cooling effect; they reflect much of the sun's rays back into space before they can warm the surface. Overall, it is thought that clouds reduce by some 20 percent the amount of sunlight hitting the Earth's surface.

Some scientists predict that an increased greenhouse effect will warm the seas, thereby causing more water to evaporate and more clouds to form—thus counteracting some greenhouse warming. Other scientists caution, though, that there is no way to predict what type of clouds will form or where they are most likely to form—both important factors in determining whether clouds will help warm or cool the Earth.

One indicator of the complexity of clouds is that most of the global climate models agree on a wide range of scenarios of global warming—*until* clouds are included in the projections. At that point, the models vary dramatically in their forecasts.

Evaluating the Models

Despite the uncertainties, these computer models of the atmosphere have proved their worth by accurately replicating conditions that occur in the real world—such phenomena as the change of seasons. The models have also been tested by inserting data from periods in Earth's history, such as the warming after the last ice age, for which scientists have solid knowledge of regional climates. If the model churns out a world that has warm and cool spots matching, for example, the regional patterns reflected in fossilized pollen grains of warm-loving and cold-loving plants, then confidence in the model is boosted.

In 1990 the Intergovernmental Panel on Climate Change, a group of several hundred eminent scientists convened by the World Meteorological Organization and the United Nations, published a detailed assessment and comparison of the most sophisticated climate models. The panel's goal was to provide a framework of knowledge upon which governments can base plans for the future. Although different models disagree sharply on some of the details of Earth's greenhouse future—some show hot spots where others have cool spots, for example—they all agree that a world with double the pre-industrial level of carbon dioxide will be a significantly warmer place.

In the Arctic, sea ice may break up and melt in response to rising temperatures. In the absence of all that white ice, which reflects sunlight and thus keeps the region cool, the dark water would absorb more solar energy, adding to any warming.

Predictions range from 3° to 8°F (1.5° to 4.5°C) warmer. A "most likely" figure for the increase in global mean temperature is 4.5°F (2.5°C) warmer.

The models also consistently predict that the Earth's polar regions, especially the Arctic, will be close to the current norm in the summer, but perhaps 14°F (8°C) warmer than the current average in winter—a drastic change that is likely to have important consequences for climate. There is general agreement, for example, that such warming will reduce the total amount of sea ice and also reduce the extent of land covered by snow in winter. These changes are expected to have an amplifying effect, further warming the globe. Surfaces covered in snow and ice reflect almost 90 percent of the sunlight that strikes them; less snow and ice means that more solar energy will be absorbed by soil and sea water and other surfaces, and thus Earth will grow even warmer.

In addition, several of the leading models, particularly the model running at the Geophysical Fluid Dynamics Laboratory in Princeton, New Jersey, predict a dramatic change in the global water cycle. Because of the increased warmth, more water will evaporate from the oceans, more clouds will form, and more precipitation will fall. The precipitation, however, will not be evenly distributed. While some regions grow wetter, others will dry out.

The Rising Tides

Most researchers also agree that greenhouse warming will cause a substantial rise in sea level. A warmer atmosphere is expected to accelerate the melting of ice and snow on land masses near the poles, increasing the runoff of fresh water into the sea. In addition, some of the warmth from the air will be transferred to the oceans. Heat causes the molecules within an object to move at greater speeds; they thus take up more space, causing the object to expand slightly. In a much warmer world, then, seawater is expected to expand a fraction of a tenth of a percent.

Seas have already risen slightly over the past hundred years, at the seemingly imperceptible rate of about half an inch per decade. That increase does not sound like a lot, but the result has already been substantial erosion of beaches and salt-water intrusion into such important agricultural regions as the Sacramento River valley in California. Now the conservative estimate is for the seas to rise three to six times faster through the coming century, reaching more than a foot and a half above current levels in the next eighty years. The rise may be dramatically higher if the warming accelerates the melting of glaciers and calving and melting of icebergs.

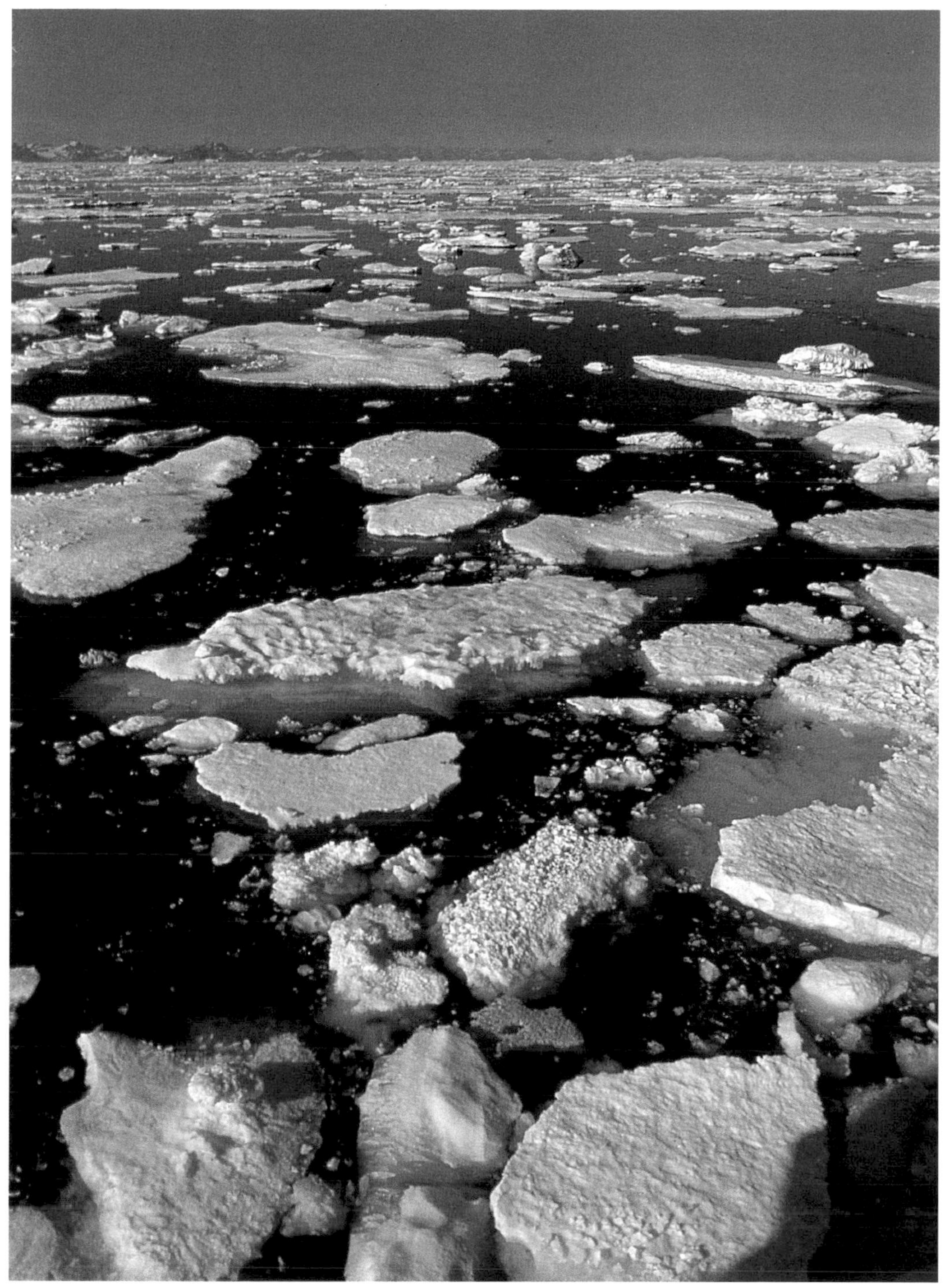

Some scientists contend, though, that any such accelerated melting will be counterbalanced as increased snowfall over Antarctica (thanks to all the extra clouds and precipitation) takes more water out of the oceans and stores it in frozen form on dry land.

A worst-case scenario, according to some researchers, concerns the West Antarctic Ice Sheet—an India-size slab of ice a mile thick. This mass of ice rests on the sea floor, like a ship aground on a submerged reef. As the sea rises and warms, it is possible that the ice could eventually become destabilized and break up, with a resulting rise in sea level of ten feet or more worldwide.

Studies are under way to check the stability of the Antarctic ice, to anticipate feedbacks that could accelerate global warming, to narrow the uncertainties in the models. Inevitably, however, no matter how refined the models become, there will be surprises in store. The West Antarctic Ice Sheet may not slip at all. But something else beyond the scope of our current knowledge may be disrupted.

Indeed, although the climate models all agree that warmer times are coming, they may give the wrong impression by implying that the world will change in a seamless, stepwise fashion in response to the steady buildup of carbon dioxide and other gases. Nature is just as likely to respond in some sudden, unpredictable way, thanks to some overlooked, variable, or misunderstood property of the natural world. The chances are very good that such surprises will have a detrimental impact on ecosystems and economies.

This possibility has been stressed by Wallace Broecker, a geochemist at Columbia University's Lamont-Doherty Geological Observatory, which nestles in the woods on the west bank of the Hudson River north of Manhattan. Broecker has documented, in ice-core and sea-floor samples, past periods when an abrupt shift in the salinity of the North Atlantic suddenly shut down the ocean currents that keep Europe warm—as suddenly as if someone turned off a faucet. Around 10,800 years ago, for example, such an event resulted in a dramatic return of ice-age conditions to Europe, followed by a swing back to warmer conditions.

It is conceivable that the increased greenhouse effect could melt enough ice to release a torrent of fresh water that would similarly reduce the salinity of the North Atlantic. Broecker summed up his thoughts in testimony he submitted to one congressional hearing: "Earth's climate does not respond in a smooth and gradual way; rather it responds in sharp jumps. . . . If this reading of the natural record is correct, then we must consider the possibility that the major responses of the system to our greenhouse provocation will come in jumps whose timing and magnitude are unpredictable. Coping with this type of change is clearly a far more serious matter than coping with a gradual warming."

Unpleasant Surprises in Store

That the natural world is full of surprises was made clear in 1985, when scientists first reported the detection of a hole over the Antarctic in the protective layer of stratospheric ozone. By that time, atmospheric chemists were in agreement that chlorofluorocarbons posed a serious threat to the ozone shield, but projections were that the ozone would be depleted gradually—perhaps 2 percent over the next hundred years. One indicator of the lack of concern was that world consumption of CFCs, after dropping slightly in the 1970s, was up to 2 billion pounds a year by 1985.

No one anticipated that certain conditions above the poles—clouds of ice crystals and a spinning vortex of winds—would create a sudden regional hole in the layer. In fact, a computer that monitored satellite scans of the ozone layer had for years been rejecting the annual appearance of a hole in the layer, ascribing the strange readings to instrument error.

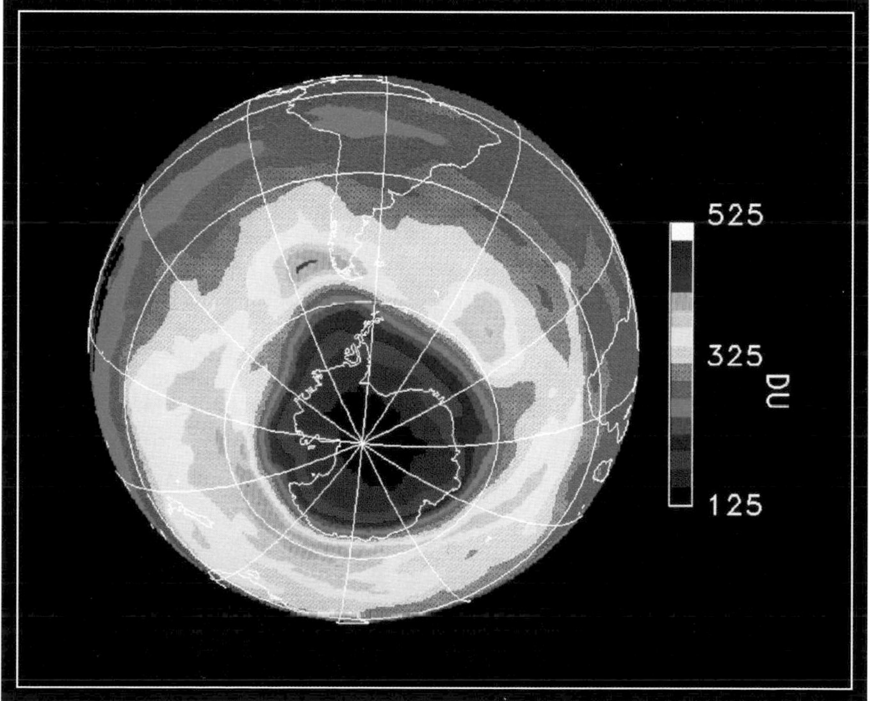

A satellite view of Earth, looking at Antarctica on October 3, 1990, shows the so-called ozone hole—a strong seasonal depletion of the region's protective shield of ozone. The purple and black areas show the greatest drop in ozone levels. Scientists had predicted that rising levels of chlorofluorocarbons (CFCs) would deplete ozone, but no one foresaw such a sharp local change. Similar unpleasant surprises could result from the buildup of greenhouse gases.

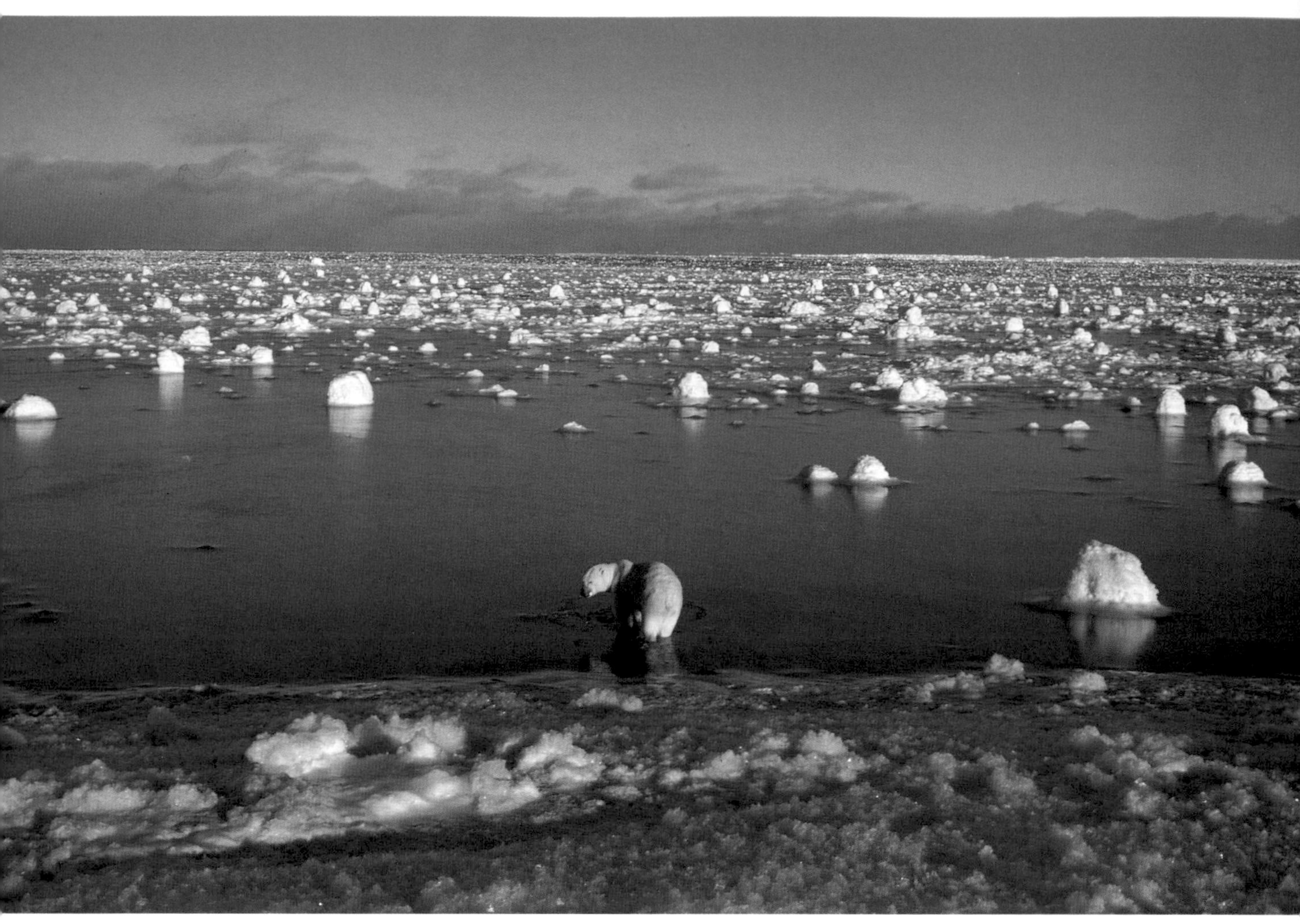

The depletion of the ozone layer and the intensifying greenhouse effect both threaten the arctic environment. Above, a polar bear glances over its shoulder in front of a landscape of sea and melting ice.

The Antarctic hole varies in size but has generally grown larger over the last decade. Ozone loss has since been found over the North Pole as well, and there has been a general thinning of the ozone layer over the mid-latitudes. Even that depletion has been found to be accelerating twice as fast as anticipated.

The appearance of the ozone hole, although not directly related to the greenhouse effect, provides a classic example of an unexpected, unpleasant twist in the dynamics of the atmosphere. And it is a delayed-reaction surprise, as well. Because it can take many years for an individual CFC molecule to migrate into the stratosphere, where it can start attacking ozone, the depletion we are seeing today is being caused by CFCs released years ago. In the meantime, human activity has pumped additional millions of tons of these chemicals into the air, and *their* effect will not be felt for years to come. We may be in for an even bigger surprise.

The lesson of CFCs can be applied directly to the looming problem of greenhouse warming. Many atmospheric scientists say we are literally taking a global gamble by modifying the atmosphere so significantly, and so quickly. Stephen Schneider, a leading climate modeler at the National Center for Atmospheric Research, calls the current situation "climate roulette." Michael Oppenheimer, a senior scientist at the Environmental Defense Fund, says: "We have an obligation to weigh the risks of inaction against the cost of action. In that regard, global warming is no different than any other problem. But global warming is novel in one respect. It brings with it the possibility of global disaster, and we have only one Earth to experiment on."

Bill McKibben, author of *The End of Nature*, recently gave a talk at Columbia University in which he encouraged the audience to make personal choices that will help soften the human impact. He described the situation this way: "If you become an environmentalist, people will say you're a radical. But that's not the case. What *is* radical is saying, 'Hey, let's double the amount of CO_2 in the atmosphere and see what happens.' That is a really radical thought. And that is exactly what we are doing."

BUSINESS AS USUAL

"Homo sapiens is perceived to stand at the top of the pyramid of life, but the pinnacle is a precarious station."

—Patrick Leahy,
U.S. Senator (b. 1940),
in a 1978 defense of
endangered species

On April 22, 1990, the twentieth anniversary of Earth Day, there were optimistic pronouncements at rallies and on television talk shows that we were witnessing the start of the "Green Decade." It looked as though the lessons of the 1980s —the medical waste on beaches, the oil slick in Prince William Sound, the endless summer of 1988, the fires in the Amazon, the hole in the ozone layer—were galvanizing a wide public commitment to changing the way the world worked. Canned tuna was made "dolphin safe." So-called "green" products flooded the marketplace. McDonald's even announced plans to eliminate its fifteen-year-old foam clamshell burger package, after a lot of urging from environmental groups and thousands of schoolchildren. The 1990s, it was predicted, would be a decade in which humanity would learn to live within its means, conserve resources, and respect the laws of nature.

On August 2, 1990, however, a conflict erupted in the oil-rich Persian Gulf that got the green decade off to a very black start. By the time Iraq retreated from Kuwait six months later, massive oil spills from shattered pipelines and storage tanks—many times greater than the spill that soiled Alaska in 1989—had spread across hundreds of square miles of the azure gulf waters. Viscous black waves lapped at the sandy shores of Kuwait and Saudi Arabia with a plopping sound, carrying the corpses of cormorants onto the beach. The shallow gulf, home to dolphins, small whales, and hundreds of species of fish and mollusks, acquired an iridescent sheen, like the puddles in front of an auto-repair shop.

Ashore, more than six hundred Kuwaiti oil wells gushed torrents of flame. Roiling black columns of smoke snaked into the sky and coalesced into a massive cloud of soot and toxic compounds that drifted hundreds of miles and coated the white desert sand with a layer of black gunk. Dozens of refineries, pipelines, tank farms, and wells in Iraq had also been hit by American and allied bombs or missiles, adding to the choking pall. All over the region, the smoke shrouded the earth in darkness, forcing drivers to use their

The Persian Gulf War got the so-called Green Decade off to a very black start. In the aftermath of the war, smoke from burning oil wells turns day to night near the Burgan Oil Field of Kuwait (preceding pages and opposite).

headlights at noon. Something like 5 million gallons of oil was burning every hour, and it was expected that it would take many months to extinguish the last of the oil fires.

The smoke and flames were perhaps a fitting epitaph for a war fought in some measure to ensure the free flow of oil. When Iraq invaded Kuwait and oil prices shot beyond $40 a barrel, there was suddenly talk in the United States of gasoline taxes and crash conservation programs. The nation seemed finally to recognize that reliance on oil, besides setting the stage for an environmental crisis, literally had the economy over a barrel.

Even before the lightning-quick land war was over, however, everything was back to business as usual. By war's end, the price of gasoline at American pumps had dropped below the price that prevailed before Iraq's invasion of Kuwait. The price of light, sweet crude dropped to $19 a barrel. In the last week of the war, news reports of the release of the Bush Administration's long-delayed National Energy Strategy were buried in the middle pages of newspapers.

Critics were quick to attack the strategy, charging that it was no strategy at all—it contained no plan for energy taxes, no new auto mileage standards, few incentives for energy efficiency. Among the few concrete actions it did propose were the opening of new offshore areas for drilling and the exploration of the oil potential of the Arctic National Wildlife Refuge. In a speech in Washington, the Secretary of Energy, Admiral James D. Watkins, explained that the policy simply reflected the wishes of the American people. As he put it, "They really do believe the Bill of Rights gave them unleaded regular for $1.06 a gallon, and they better get it or, by God, they'll get the bums out of office."

The Status Quo: Trouble down the Road

Given the likelihood that the nation, and to a large extent the world, will continue—at least for the time being—with business as usual, it is useful to examine just where various scientific studies say this approach to the greenhouse problem would take us in the coming century.

In this "business-as-usual" world, any response to the creeping threat of global warming is put off until tomorrow, like an overdue paint job for a slowly rusting bridge, or a repeatedly canceled checkup for that odd-looking birthmark. The industrialized nations continue along the path of unfettered growth, while developing nations follow right in the tracks of the global powerhouses, eager to become powerhouses themselves.

The business-as-usual world includes a China that doubles its use of coal

By the year 2000, China plans to double its use of coal—and thus greatly increase emissions of heat-trapping carbon dioxide. Just as happened in the West, industrial growth is being put far ahead of environmental concerns, as is evident at this steel and iron complex in the Nei Mongol region (formerly Inner Mongolia).

in ten years. It is a world, remember, in which the number of automobiles doubles. It is a world in which tropical rain forests, the greatest reservoirs of biological diversity, are largely replaced by scarred and eroded hillsides and weed-choked pasture. It is a world with a steadily rising human population, which is not expected to level off until it has reached 10 billion people sometime around 2060. And many of those people look longingly at the level of prosperity in the industrialized West—a prosperity nourished by the use of three gallons of oil per person per day. In such a world, the atmospheric impact of greenhouse gases doubles very, very quickly.

The computers say that such a world will, on average, be 4.5°F (2.5°C) hotter; seas will be 1.5 feet higher; there may be more precipitation in the Arctic and less in the hearts of continents. Such bare-bones predictions, however, are not very meaningful to most people. After all, the tides rise and fall several feet twice a day in coastal areas—and thirty or forty feet in places like

If present trends continue, the number of cars on Earth is expected to double by 2030, greatly increasing emissions of carbon dioxide. In this traffic jam, on Bangkok's Rama IV Road, a mass of vehicles sits idling.

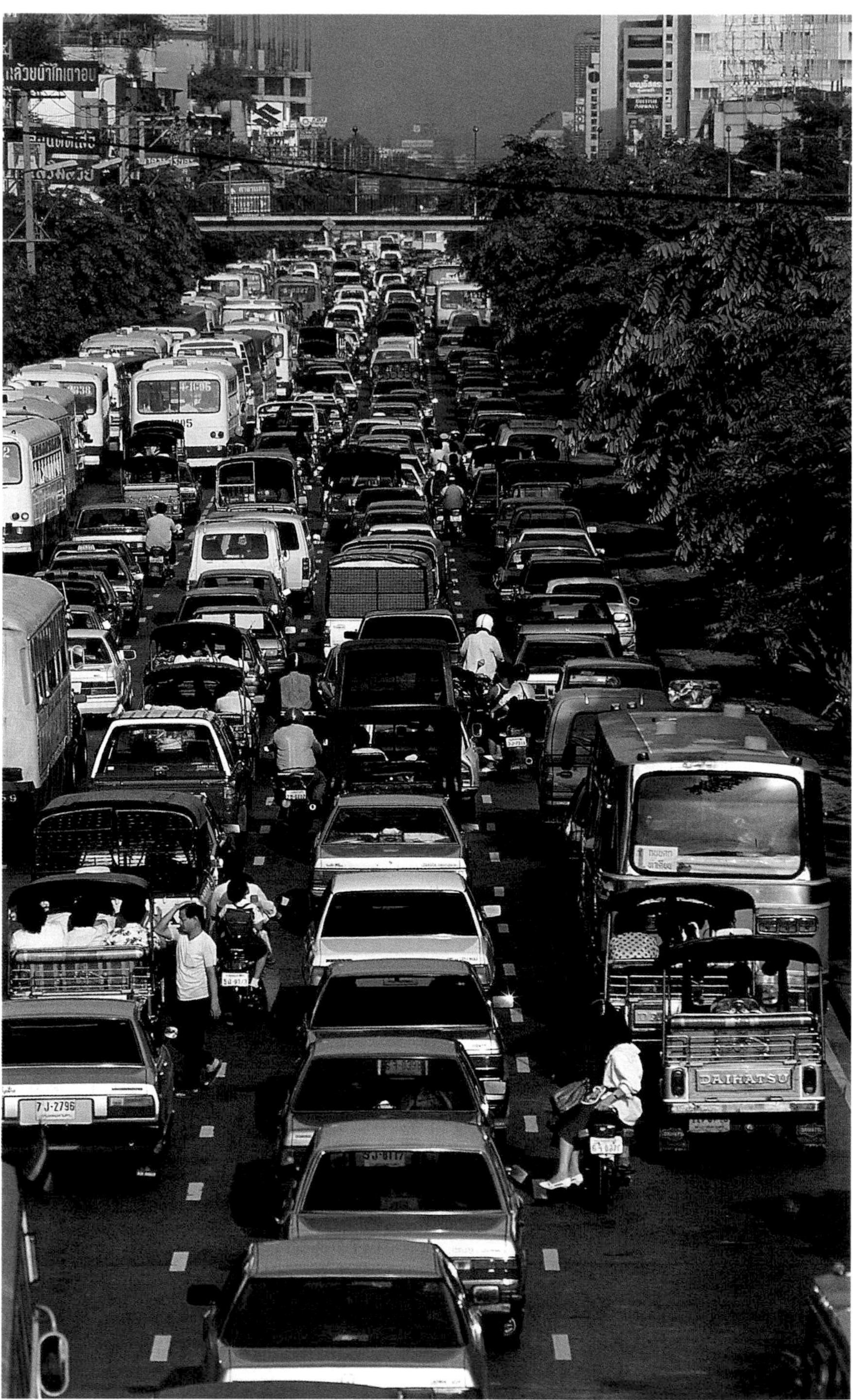

Nova Scotia and Darwin, Australia. What's another foot or two? And on almost any day anywhere in the world the temperature varies from noon to midnight by at least a few degrees. Most people want to know what is in it for them: What does all of this mean to rice farmers in the Philippines; ranchers in Kansas; or retirees in Miami Beach? This kind of close-focus prediction is the hardest to make, but it is worthwhile taking a closer look, even though the details are sketchy and represent only possibilities.

Field studies and more fine-tuned computer analyses are being used to fill in the possible local impacts of these global changes. It is such regional assessments that most concern planners and politicians, who, after all, have local constituencies. Humans have a natural tendency to react most strongly to problems that are closest to home.

Feeling the Heat

That projected rise in global mean temperature will not be discerned, probably, in the course of everyday life; it is unlikely that someone standing on a street corner in Dallas or kayaking on a river in Labrador will take out a thermometer and note that the temperature is 4.5°F (2.5°C) higher than it was fifty years ago. The most noticeable change, according to many climatologists, will be the increased likelihood of an unusually hot day or dry spell. "January thaws" may occur sporadically throughout a winter. Summer heat

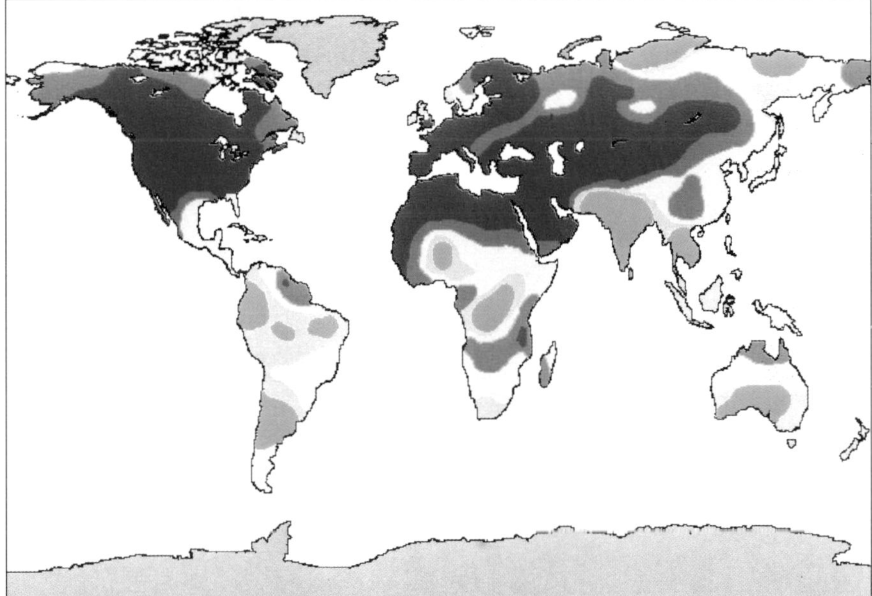

- 30–60 percent drier
- 20 percent drier
- 10 percent drier
- No change
- 20–100 percent wetter
- Not calculated

Patterns of rainfall and drought are expected to shift sharply as greenhouse gases continue to build. This illustration, derived from a computer projection, shows that broad regions of the United States, Europe, and Asia may become far drier than they are today.

waves may hit North America with greater frequency. For example, Dallas today has an average of nineteen days each summer that top 100°F (37.8°C). Researchers at the Goddard Institute say that, in a world with doubled carbon dioxide levels, Dallas may experience something like seventy-eight days with such scorching readings. Washington, D.C., today has an average of only one day each summer that hot; doubled CO_2 levels will bring nearly two weeks of 100°F heat to the nation's capital. In many ways, it is not the gradual temperature rise, but the increased frequency of weather extremes that will be most bothersome.

Because the anticipated climate change will not be smooth and uniform, prospects for world agriculture are uncertain. There are sure to be places where agriculture will benefit: Warmer weather would extend growing seasons and open up new territory for farming in such places as the Soviet Union and Canada. Moreover, increased carbon dioxide levels may greatly boost the productivity of plants, because more carbon dioxide means more photosynthesis. Also, some plants use water more efficiently when grown in an atmosphere rich in carbon dioxide. Crops such as wheat, however, need more than just carbon dioxide to grow. They also need the right soils and adequate water. The optimum climate for wheat growing may move northward in Canada, but the rich soils of the plains will not be there to sustain wheat cultivation.

A factor that may offset any agricultural benefits from rising CO_2 levels is the impact of higher temperatures on the planet's water cycle. Although precipitation is expected to increase, much of the additional precipitation may fall near the poles, benefiting few farmers. Precipitation patterns may shift sharply. The climate model at the Geophysical Fluid Dynamics Laboratory predicts that India will have much more rain, while the midwestern United States will be 30 to 60 percent drier in summer than today. Some arid

Increase in Temperature

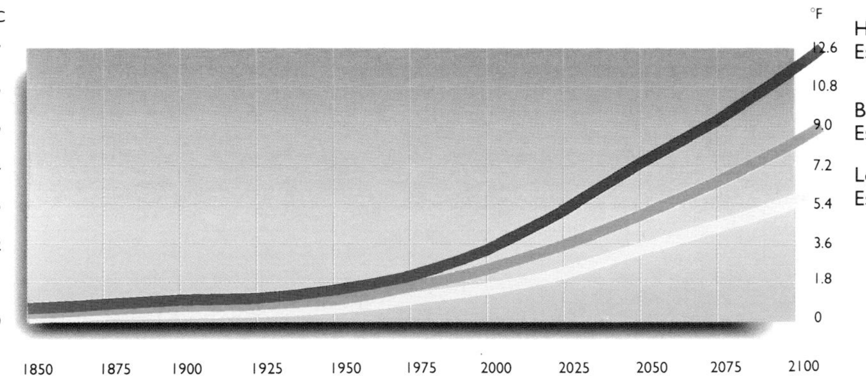

If greenhouse gases continue to increase roughly at current rates—the so-called "business as usual" scenario—global temperatures could rise more than 8°F (4.5°C) by the end of the next century, according to the best estimate.

As the greenhouse effect intensifies, weather extremes—for example, heat waves—are likely to occur more frequently. Each summer New York City has typically had just 15 days over 90°F (32°C). By early next century, it may have 48 days each summer with temperatures that hot. Here, children cool off in a fountain on a scorching day in Manhattan.

regions, such as southern California and Morocco, will have less precipitation in winter—the season when such areas receive most of their precipitation. Areas that rely on melting mountain snows for year-round irrigation, such as California's San Joaquin valley, may find themselves in trouble. Warmer temperatures will cause more winter precipitation to fall as rain, leaving less snow to provide water through the spring and summer. Moreover, even in areas where there is no diminution of rainfall, higher temperatures may lead to drier soils; and much plant stress is related directly to temperature, not to moisture.

Such changes could further destabilize already volatile regions of the world where nations are fighting over water. Egypt and the Sudan, for example, both draw much of their water from the north-flowing Nile. Sudan has been trying to divert a larger share of the river's water; but downstream, Egypt is in the middle of a population explosion and needs more water than ever. A string of droughts in the Sudan could lead to water wars.

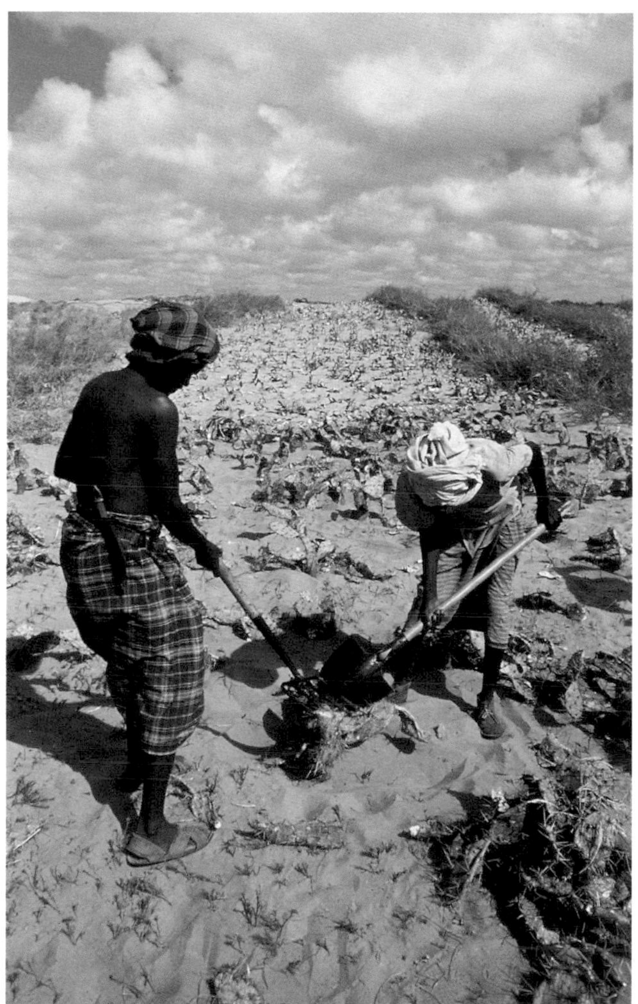

Worldwide, water supplies may be disrupted as patterns of precipitation shift. Already, nations around the globe are having trouble finding sufficient water to grow crops and slake thirsts. Opposite, a drought bakes the dry bottom of France's Sombernon Reservoir. Left, Somali workers plant cactus to preserve what little soil they have. Below, a well in Somalia has been progressively excavated as the water table has dropped.

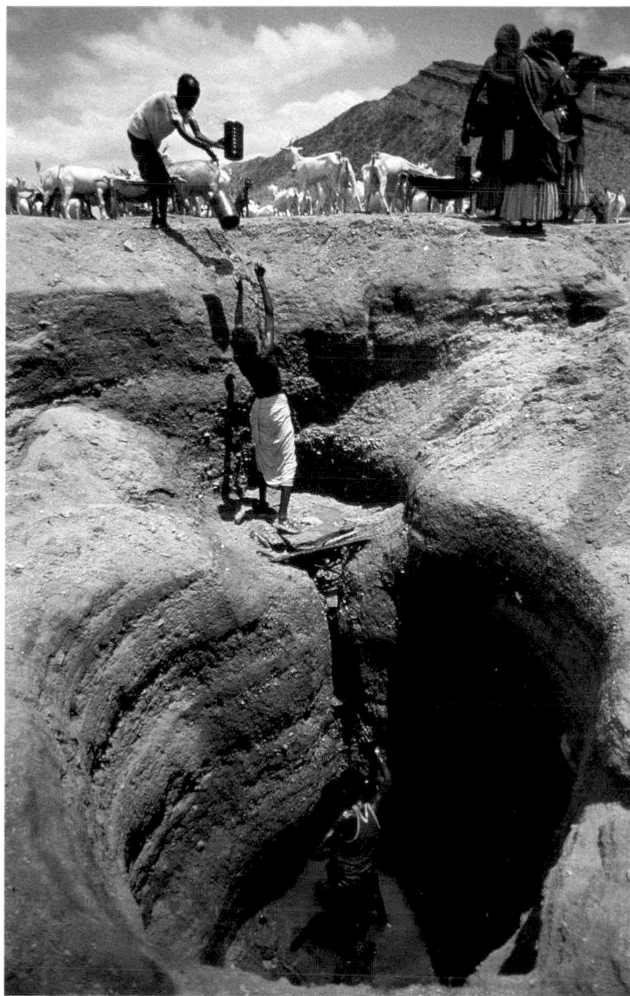

Islands at Risk

Perhaps the most straightforward projections of what global warming will bring in coming decades are related to rising seas. A foot and a half rise does not sound like much—unless you live in a place that just barely pokes above the ocean.

At the First International Conference on the Changing Atmosphere, I met Hussein Manikfan, the permanent representative to the United Nations from the Republic of Maldives, who made this fact clear. At first it seemed odd to find a representative from the Maldives at the meeting. Most of the discussions centered on devising strategies to curb emissions of carbon dioxide and other heat-trapping gases from automobiles, power plants, and the burning of tropical forests. The Maldives, a sprinkling of 1,190 coral islets in the Indian Ocean southwest of Sri Lanka, have no tropical forests, hardly any automobiles, and little industry beyond the canning of bonito. I spoke awhile with Manikfan. Why was he in Toronto? "To find out how much longer my country will exist," was his simple reply.

Manikfan is worried because few of the islands have any point more than six feet above sea level. Even now, in strong storms many of the atolls are awash. The fear is that Manikfan's nation—which has its own language and alphabet, and a tradition of independence dating back thousands of years—

If seas rise and warm as a result of global warming, the habitat and food supply of coastal species, such as this green turtle, may be threatened.

The 200,000 people living on the 1,190 islands of the Republic of Maldives may be the first humans to feel the direct impact of global warming. Few spots in this Indian Ocean archipelago, home to a 2,000-year-old culture, rise more than six feet above sea level. Even a small rise in sea level will increase damage from flooding and storms.

may have to be abandoned altogether, as if it were a slowly sinking ship. Such a calamity would mark the first time in recorded history that an entire nation would have to relocate because of an environmental problem.

The problem of vulnerable island nations extends from the Indian Ocean to the Caribbean and South Pacific. Although the full impact of rising seas will not be felt for decades, island nations are already planning ahead. The greenhouse effect was a central topic of discussion at a recent meeting of the South Pacific Forum, hosted by the Micronesian nation Kiribati. This nation, like other Pacific atolls, has almost no land more than six feet above sea level. "You only have to be here to see that you don't need much of a rise in the oceans and you're talking about 'Goodbye, Kiribati,'" Australian Prime Minister Bob Hawke said at the meeting. Hawke pledged $4.8 million (US) to establish a network of scientific stations around the Pacific to track changes in sea level, atmospheric pressure, and temperature. Hawke also promised representatives from tiny Tuvalu and Kiribati that Australia would consider resettling any eco-refugees.

Coastal Encroachment

The coastal regions of the continents may also be in harm's way, particularly towns or cities built on barrier islands and the fertile flat plains that typically surround river deltas. Bangladesh, dominated by the Ganges-Brahmaputra-Meghna delta, is the classic case, according to Robert Buddemeier, a geological chemist who is studying the impact of sea-level rise on coasts and islands around the world. "It's massively populated, achingly poor, and something like a sixth of the country is going to go away," he says.

In Bangladesh, the threat of disaster from ocean surges is compounded by the threat of floods when upland rainstorms swell the rivers. This problem has been aggravated by deforestation of the foothills of the Himalayas, where the river system has its headwaters. Moreover, the impact on river delta regions such as coastal Bangladesh may be amplified because these areas are already undergoing natural subsidence—sinking as the water is rising. The Nile Delta of Egypt, where 14 percent of that country's population dwells, and which produces 14 percent of its gross national product, is similarly threatened. "You're looking at an unprecedented refugee problem," says Buddemeier. "In the past, people have run away from famine or oppression. But they've never been physically displaced from a country because a large part of it has disappeared."

Closer to home, rising sea levels will spell big trouble for coastal communities. Stephen Leatherman, director of the Laboratory for Coastal Research

Bangladesh has already experienced devastating floods along its crowded coast and river deltas. At right, the city of Dacca recovers from a disastrous flood in 1988. Rising sea levels could force millions of people to abandon their land and seek a new home.

at the University of Maryland, has calculated that a one-foot rise in sea level will push Florida's high-tide mark inland by 200 to 1,000 feet. Louisiana's shorelines will move inland several *miles*. One Environmental Protection Agency study assessed the impact of a three-foot sea level rise—somewhat of an extreme scenario—on Miami. This city is nearly surrounded by water, with the swampy Everglades just to the west, the Atlantic to the east, and porous limestone—one of the world's most permeable aquifers—underneath. According to researchers, a dike would have to be sunk 150 feet deep into the earth to prevent water from welling up into the city as seas rise. Washington, D.C., built on what was once a swamp, is also very vulnerable.

Areas such as Galveston, Texas, which are located along shores protected by sandy barrier beaches, will be increasingly threatened by storm surges. Leatherman has calculated that a two-foot rise would greatly increase the impact of storms. A moderately intense hurricane—the kind that occurs about once every decade—would have the destructive impact of the type of storm that occurs once a century. A computer model of storm surges shows that Galveston would be completely underwater in such a storm, with waves rushing down the streets of Texas City, which is normally protected by the offshore islands. And Galveston is typical of a whole range of resort areas on the East and Gulf coasts, from the Hamptons of Long Island to Key West.

In cities such as Venice, Italy (right), which is already chronically troubled by floods, global warming may overwhelm efforts to adapt to rising waters.

The Gulf Coast of the United States is occasionally in the track of hurricanes such as Alicia, which flooded the streets of Galveston, Texas, in 1983 (left). One scientist projects that warmer sea temperatures will spawn hurricanes in the next century that will be 40 to 50 percent more powerful than the most potent storms of the past 50 years.

Miami Beach, Florida, is surrounded by water—and thus very vulnerable to hurricanes and rising seas. Higher sea levels will increase the likelihood that a hurricane will do severe damage.

Even as coastal settlements become more vulnerable to damage from moderate storms, the intensity of hurricanes may increase, according to the work of Kerry Emanuel, a meteorologist at the Massachusetts Institute of Technology. Hurricane intensity is linked to the temperature of the sea surface. According to Emanuel's models, if the sea warms to predicted levels, the most intense hurricanes fifty years from now will be 40 to 50 percent more powerful than the most intense hurricanes of the past fifty years.

Wilderness and Wildlife

The impact of global warming will not only be felt by humans; it will also doubtlessly disrupt the planet's remaining pockets of wilderness. Forests, for example, tend to thrive in regions where a particular set of conditions pre-

If climate patterns rapidly shift, tree species that thrive only in certain conditions may be imperiled. Above, a Tennessee forest bursts into fall colors

Opposite, Hurricane Gilbert batters palm trees along the Texas coast in 1988.

vails: the right soil type, moisture, temperature, and the like. Although individual trees are rooted in place, a forest is generally able to migrate 15 to 25 miles in a century as seeds are transported by wind, birds, or mammals. The bands of cold-loving spruce and temperate hardwoods such as hemlock and maple have thus regularly shifted north and south hundreds of miles across North America, keeping pace with the advance and retreat of ice-age glaciers. But stands of these and other trees today may be unable to shift quickly enough to keep up as the pace of change accelerates due to the buildup of greenhouse gases.

The change in temperatures accompanying the expected rise in greenhouse gases will be about one-half degree F (.3°C) per decade. This could shift

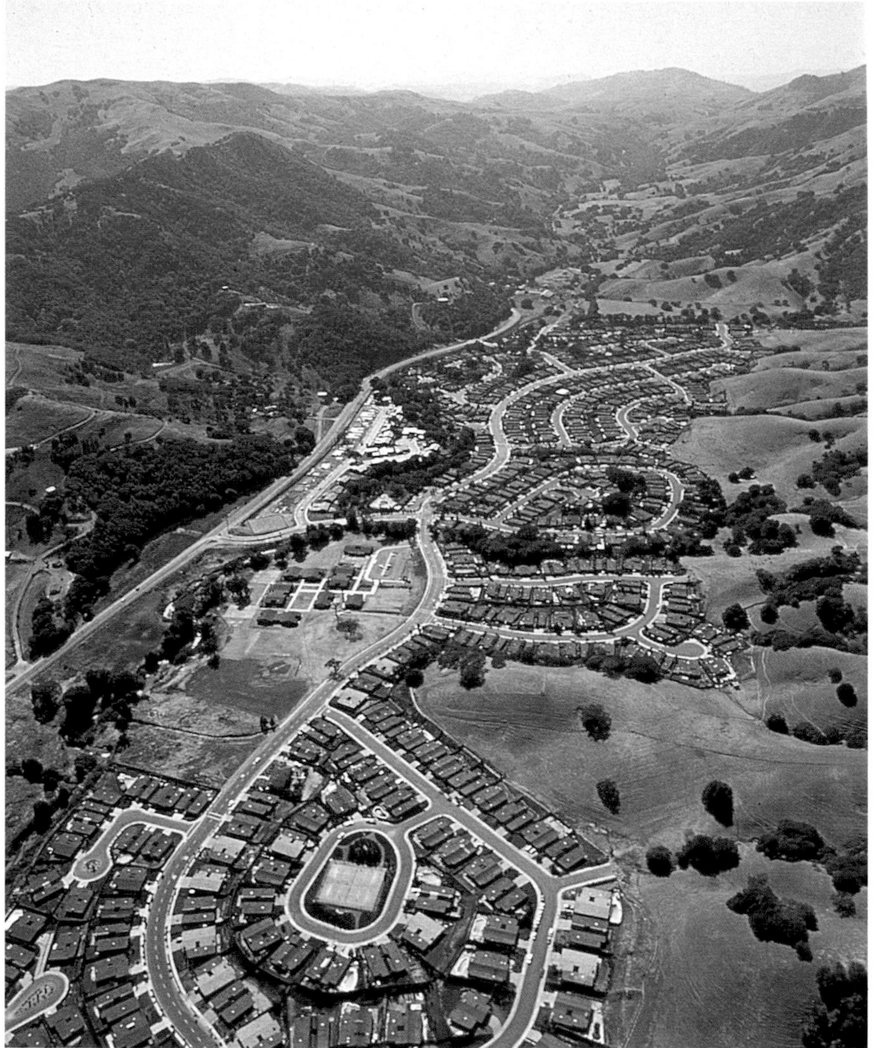

In areas such as southern California (left), housing and other development have isolated pockets of wilderness. Global warming may doom such biological fragments if species are unable to relocate along with changing climate patterns.

the climate zone favoring a particular tree species about 20 miles toward the poles every ten years. That is much faster than a forest can migrate. In other words, a forest may be left behind. When that happens, a forest dies.

Another problem facing ecosystems such as forests and wetlands is that those able to migrate fast enough to keep pace with a change in conditions may find the route blocked. Wetlands are increasingly hemmed in by such barriers as condominium communities, berms, and jetties and dredged harbors. Normally, if the sea level were to rise, the grasses and mussel beds and

Many coastal wetlands, such as Florida's Everglades National Park (above), are hemmed in by development and thus cannot shift inland if climate or sea levels change. Wetlands are a vital biological factory, supporting diverse coastal species, including the manatee (right), whose numbers are already threatened.

The Kirtland's warbler of Michigan (above) and the Florida panther (left) are only two of many animals with limited ranges that give them no route of escape if rapid climate shifts occur.

barnacles and mangroves would simply move further inland as their seeds or spawn drifted and settled on new terrain. But in many places around the United States and other parts of the world, that is no longer possible. Humans are in the way.

In the United States, most forests exist now in patches and fragments surrounded by development, and migration is difficult. National parks that harbor unique communities of plants and animals are likewise hemmed in and thus extremely vulnerable to sudden shifts in climate. When a forest or a marsh dies, so do its inhabitants. Just one such victim could be the already imperiled Florida panther.

Daniel Botkin of the University of California, Santa Barbara, has spent years studying the Kirtland's warbler, which nests only in the jack pines growing on a patch of sandy soils in Michigan. This bird may be particularly vulnerable to a sudden climate shift, in that it will have no place to migrate if the conditions peculiar to its home range change sufficiently to dry out the pines. (The warbler would also be threatened in its winter retreat, the Bahamas, which will be vulnerable to sea-level rise.) Botkin equates the warbler with the canaries that miners used to carry into their tunnels to give advance warning that the passages were filling with suffocating gases. If the warbler disappears, that will be a small alarm bell signaling that climate change is under way.

Around the world there are many such species, already stressed by the expansion of human populations and the long reach of man-made pollution. The rate of extinction of species is estimated to be higher now than it has been since the dinosaurs ended their 150-million-year reign. Some biological alarm bells are ringing, others already forever silenced. There is just one species on the planet that can respond to this tragic chorus.

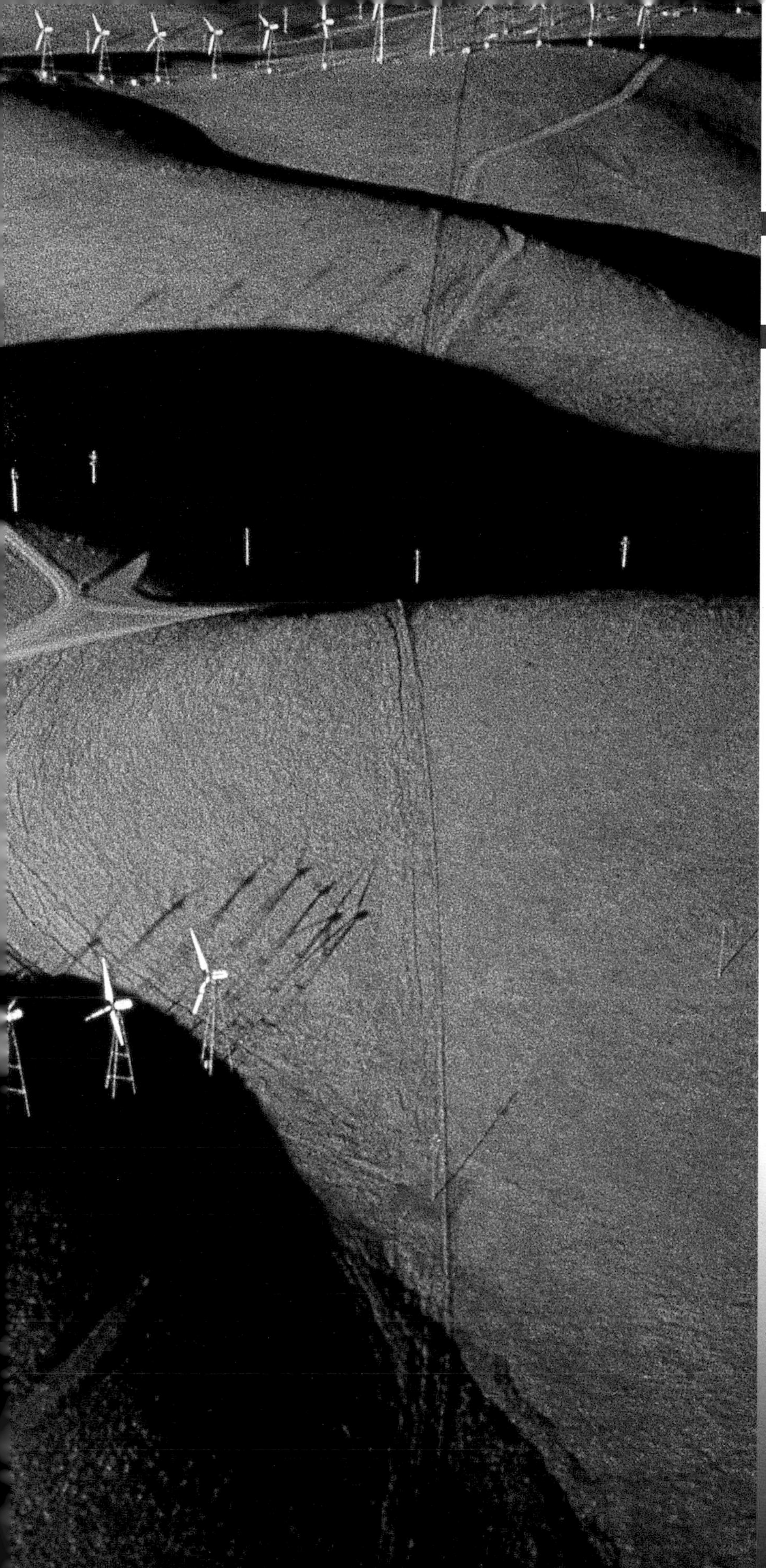

CHOOSING OUR FATE

"Man, in the words of one astute biologist, 'is . . . faced with the problem of escaping from his own ingenuity.' "

—Loren Eiseley, anthropologist and essayist (1907–1977), *The Firmament of Time*

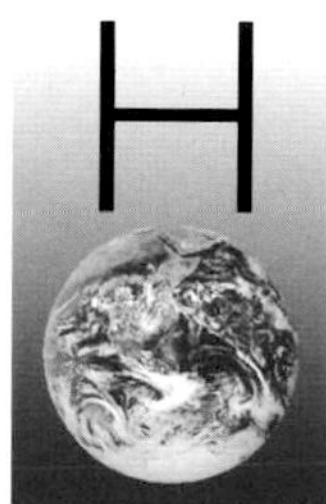

Human beings have proved to be remarkably adaptable, perhaps because the species came of age in the Pleistocene, a time of continual change. The long cycles of ice and warmth sculpted the landscape, influenced global climate, and caused the seas to rise and fall; so too did these cycles leave a mark on us. We are the antithesis of the Kirtland's warbler, which is constrained to live on a tiny patch of territory within a narrow range of conditions. We are by nature omnivorous opportunists—able to live on almost any food and to generate our own microclimate with clothing and shelter and fire. Humans, even with minimal technology, have found ways to survive from the poles to the equator. Some kneel all day by a hole in the Arctic ice, waiting to harpoon a sea lion, while others harvest manioc in the steaming jungles of the Amazon. Humans can raise goats 18,000 feet up in the Himalayas or spear fish 50 feet beneath the surface of a Polynesian lagoon.

A few economists, scientists, and planners look at the historical record and conclude that our ingenuity will get us through any coming climate change, and that the immediate cost of preventing—or at least slowing—any human-caused warming is unacceptably high. Moreover, they say, there is always the possibility that the models are wrong, and that the world is actually going to warm only moderately, or not at all. More research is needed before costly changes are made. Much more research.

Some say there is no need to worry because there will always be a technological fix. We can fertilize the ocean around Antarctica, for instance, and vast plankton blooms will pull excess carbon dioxide from the air. We can blast CFCs from the sky with specially tuned lasers. We can fill the stratosphere with planeloads of sulfur dioxide, which will form tiny droplets of sulfuric acid that will reflect away excess sunlight and counter the warming.

Charting a Course of Action

Given our current lack of understanding of the global system, however, many scientists feel that the last thing we should do is to add another variable to the

Currently, more energy escapes through the windows of American homes than flows through the Alaska pipeline. Energy-efficient designs can cut oil or gas bills and also cut emissions of carbon dioxide. Opposite, a house in Bridgehampton, New York, was designed by Preston Phillips to catch the sun's energy.

Windmills for generating electricity stand on ridgetops in California's gusty Altamont Pass (preceding pages). Efforts to harness the power of the sun and wind may help reduce dependence on fossil fuels—and thus the growth of carbon dioxide emissions.

equation. More nasty surprises would surely be in store. A more sensible course, according to a wide array of experts, is to take reasonable actions now to stem the buildup of greenhouse gases—particularly actions that are beneficial for other reasons. As another decade passes, if the evidence for global warming becomes more clear-cut, more dramatic actions can be taken to reduce levels of these gases.

The authors of an important 1991 National Academy of Sciences study concluded, "Despite the great uncertainties, greenhouse warming is a potential threat sufficient to justify action now." In part, the significance of this report arose from the wide range of viewpoints of members of the drafting committee—people ranging from the head of research for General Motors Corporation to the vice president of the World Resources Institute.

Scientists, diplomats, and politicians have been meeting with increasing frequency to discuss options to fight global warming. They have come from all around the globe, recognizing that the atmosphere is a global commons—something like the town green in a New England village or the single well at a desert oasis—a resource that serves everyone and must be maintained by everyone.

A Hopeful Precedent

One precedent has raised confidence that international action to protect the atmosphere is possible. This was the 1987 Montreal Protocol, in which 47 nations agreed to sharp cuts in the production of ozone-depleting CFCs. At the time, the main concern was the assault by CFCs on the protective stratospheric ozone shield, not their contribution to the greenhouse effect. Since then, however, there has been growing evidence of the two-pronged problem posed by the long-lived chemicals, and most of the nations party to the 1987 accord agreed in 1990 to a total ban on production of CFCs.

Of course, eliminating a single class of synthetic chemicals is a relatively simple task. Substitutes for CFCs are already being developed; what is more, these destructive compounds are produced by only a few dozen companies worldwide. As Pieter Winsemius, a former minister of the environment for the Netherlands, explains, "You can put them all in one room, you can talk to them. But you can't do that with the producers of carbon dioxide—for example, all the world's utilities and industries." Gases such as carbon dioxide and methane are by-products of the processes at the heart of modern civilization: industry, transportation, power generation, and agriculture.

Even so, researchers and policy makers see numerous opportunities for

Modern society is dependent on electric power, as this maze of power and telephone lines in a suburb of Washington, D.C., illustrates. Much can be done to reduce the greenhouse threat by using electricity more efficiently and finding non-polluting sources of electric power.

cuts in greenhouse gases. There is broad agreement that the industrialized nations will need to take the lead in slowing the growth of emissions. Three-fourths of today's output of greenhouse gases comes from these countries, even though they have just one-fifth of the planet's human population.

The United States: A Case Study

First and simplest, the industrialized nations can adopt domestic policies that encourage greater energy efficiency. There is tremendous room for improvement. Take, for example, the situation in the United States. For every unit of energy consumed, the United States produces only half the amount of goods that western Germany or Japan produces. There is reason to believe this gap can be narrowed. Partly because of the oil crises of the 1970s, the American economy found a way to grow 40 percent from 1973 to 1988 with *no* increase in energy use. The lesson is that it is possible to have healthy economic growth without a relentless rise in the use of fossil fuels.

One way that American industry can help close that gap is to watch for wasted energy in the workplace. For example, most American factories use electric pumps and motors that whir away at high speed and are modulated

A wall of recycled aluminum cans at an Alcoa recycling center in Tennessee attests to the potential energy savings in such programs. Recycling just one can eliminates the need to smelt new aluminum and saves the energy equivalent of half a can of gasoline.

with inefficient brakes or valves. Lighting is also notoriously inefficient in most plants. If industry were to upgrade to variable-speed motors and less wasteful lighting, a substantial cut could be made in the amount of electricity used (and thus in fossil fuel use and greenhouse gas emissions back at the power plant). The initial investment in new equipment would quickly pay for itself in lower energy bills.

American homes and offices are just as wasteful of energy. If we were to spend as much on insulating buildings as we now do on assuring the flow of oil from the Middle East, we would not need the oil from the Middle East, according to Amory Lovins, founder of the Rocky Mountain Institute, a Colorado think tank for energy-efficiency research. More energy escapes through the poorly insulated windows of American homes than flows through the Alaska oil pipeline. So-called "superwindows" insulate as well as twelve panes of glass, keeping homes cooler in summer and warmer in winter. The added cost of the new technology is quickly recouped in savings on heating bills.

The biggest consumer of electricity in the typical home is the refrigerator, and models exist that are several times as efficient as the current average. The incandescent light bulb also is notoriously inefficient, with most of the electricity used transformed into heat, and only a small fraction converted into light. The heat from such bulbs, incidentally, adds to air-conditioning costs in

the summer. Environmental groups have shown how utilities can benefit in the long run by helping their customers invest in high-efficiency alternatives that consume less power. When demand is cut, the companies are able to spend less to fuel their existing plants and avoid the need for expensive new power plants.

One recent example of this thinking is a program in which several big electric utilities are giving away hundreds of thousands of energy-saving compact fluorescent bulbs. These bulbs screw into standard sockets and throw the same amount of light as an incandescent bulb, but last ten times as long and consume just one-fourth as much power.

Common Sense and Scientific Research

Simple, sensible practices such as recycling and waste reduction also save energy. These practices also reduce the demand on the nation's overflowing landfills. As scientists at the Rocky Mountain Institute have noted, if you discard an aluminum can, you waste as much energy as you would if you half-filled the can with gasoline and poured it on the ground. The institute has calculated that in 1988 alone, the recycling of aluminum cans in the United States saved more than 11 billion kilowatt hours of electricity—enough power to satisfy the needs of New York City for six months.

Energy use—and greenhouse emissions—can also be cut by boosting funding for basic research into new efficient technologies. Japan has taken the lead in this area, establishing a $40-million research institute devoted to developing technologies that solve environmental problems. Dozens of companies, including Toyota, Hitachi, and Nippon Steel, are contributing money and researchers to the project. Some analysts are predicting that environmental technologies will boom in the 1990s. The United States would do well to get into this field, these analysts say, if only to increase international competitiveness.

Oil Dependency

In the United States, much can be done to curb the national appetite for oil, which far exceeds that of any other country, and is mainly accounted for by transportation. More than 60 percent of the 17 million barrels of oil burned here each day—and thus transformed into carbon dioxide and a host of pollutants—goes into vehicles. The great majority of that goes into the tanks of automobiles and trucks.

Part of that runaway consumption is a function of the long distances many Americans drive. The total number of miles driven in this country has risen steadily, year after year. It is projected that the number of miles driven will double by 2020, if present trends continue. The other side of the problem is the tens of millions of inefficient vehicles on American roads. Some are simply old and out of date, but many are fresh from the showroom floor, where gas-guzzling engines and long wheelbases have recently come back into style.

Leaving Cars at Home

In the transportation sector of the economy, the biggest greenhouse reduction of all would come if more people simply left their cars at home and rode subways, buses, bicycles, or trains. Along with boosting support for mass transit, governments can set higher standards for gas mileage or increase the cost of fuels with higher taxes.

In the United States, the average federal and state taxes on a gallon of gas total around twenty-five cents. In countries from Japan to France, taxes range from $1.40 to $2.30 a gallon—a powerful incentive to consumers to buy efficient vehicles. Through the 1980s, the United States government resisted raising the gasoline tax. The same held true for gas-mileage standards for vehicles. The current required average for a car company's new models is

27.5 miles per gallon, even though eight auto manufacturers have developed prototype vehicles that get anywhere from 64 to 112 miles per gallon.

In the 1990s, though, there is a growing chorus of voices calling for higher mileage standards and higher gasoline taxes, to benefit both the economy and the environment. The voices are not just those of environmentalists. Recently, a former vice president of Gulf Oil, Ben C. Ball, Jr., wrote in the *New York Times,* "If we are serious about conserving oil, we should join the rest of the industrialized world and raise petroleum taxes, not by pennies but by dollars per gallon. . . . Conservation would be an economic choice; imports would be reduced and pollution would lessen." Moreover, Ball said, the U.S. government still spends billions of dollars annually on a maze of programs and subsidies that keep energy prices artificially low and energy consumption high. For example, there is plenty of funding for highways, but little for mass transit; current housing policies encourage urban sprawl (and thus more car miles traveled); more federal support goes to air and truck transport than to energy-efficient railways.

Energy taxes are no panacea. It is hard to predict, for example, how great a reduction in carbon dioxide emissions would be produced by a particular tax level. There are, however, additional measures that industrialized nations can take to cut emissions of heat-trapping gases. One idea is to pro-

Big energy savings can be had if more people opt for mass transportation, higher-mileage cars, or—where possible—human-powered vehicles. Here, bicyclists take over Manhattan's Fifth Avenue on a day when cars are banned from the street.

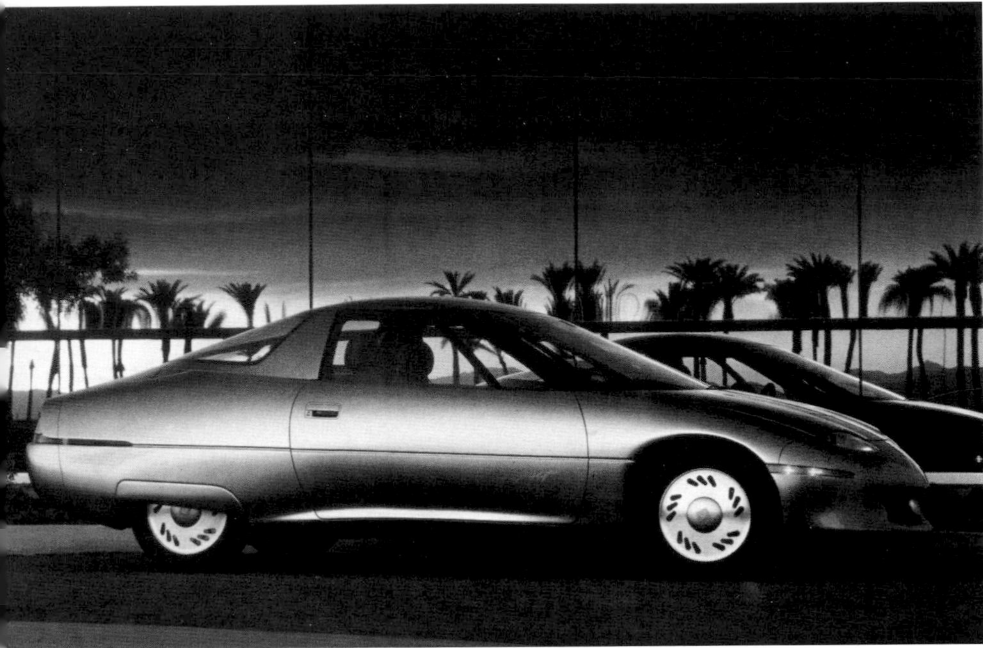

General Motors displays a prototype for the electric-powered Impact (left). In 1991 General Motors announced that it would convert an assembly line to the production of electric cars.

mote alternative fuels that produce less carbon dioxide than does coal or oil. When burned, natural gas produces 28 percent less carbon dioxide for each unit of energy released than does oil, and 44 percent less than coal. The fuel has an octane rating of 130, far above that of gasoline. Also, because natural gas burns cleanly, engines using it have required far less maintenance. It is also cheaper than gasoline. Natural gas is already proving itself in regional experiments. In British Columbia, tens of thousands of vehicles are running smoothly on natural gas. In the United States, the United Parcel Service has switched hundreds of its trucks over to natural gas and may eventually shift its entire 80,000-vehicle fleet.

Solar Solutions

Nations can also promote development of existing clean energy sources, such as solar power and wind power. Various solar technologies are showing great promise: both the use of vast arrays of mirrors to generate steam and the use of photovoltaic cells to convert sunlight directly into electricity. By 1994, Luz International—a Los Angeles firm that is the world's leading producer of solar electric power—will be generating 680 megawatts at its facility in the Mojave Desert. This is two-thirds the output of a typical nuclear plant. In the United

The electric-powered LA301 sedan (right) was developed by the European company CleanAir with support from the smog-plagued city of Los Angeles. It is expected to be the first production-line electric car sold in the United States.

States, the cost of converting sunlight to electric power has steadily dropped despite more than a decade of cuts in government funding for solar-power research. In Germany and Japan, where research funding has not flagged, the technology is even more advanced.

Wind power is showing its potential in many spots around the world. (Wind is an indirect manifestation of solar energy—masses of air set in motion by the heat of the sun.) In California, 17,000 high-tech windmills have been built, from the Altamont pass in the north to Palm Springs in the south. These windmills generate enough energy to meet the needs of all the homes in San Francisco.

What Role for Nuclear Power?

Increased use of nuclear power may someday play a role in cutting emissions of greenhouse gases—after all, the fission of radioactive materials releases no carbon dioxide at all. Proponents point to successes in France, where more than 60 percent of electric power is generated by nuclear plants. Some new reactor designs on the drawing board are considered inherently safer than existing types.

Nonetheless, many obstacles must be overcome before nuclear power

Various strategies, from photovoltaic cells to solar steam-powered turbines, are being studied for converting solar energy into electric power. Above, the Japanese "Sunshine" project employs different types of mirror arrays.

can be considered safe and clean. There is still no solution to the problem of how to handle and dispose of the deadly, long-lived radioactive waste that such plants produce. Holding pools at nuclear power plants around the United States are quickly filling to capacity with high-level radioactive waste, and the country still has not agreed on where to store the waste permanently. Meanwhile, accidents at Three Mile Island, Chernobyl, and, most recently, near Kyoto, Japan, have heightened public concern about the potential hazards of nuclear power. Moreover, the proliferation of nuclear fuels may increase the rate at which nuclear weaponry proliferates.

Thinking Globally

One of the most effective ways for the world's more advanced nations to slow the growth of the greenhouse effect is to help the developing world

In California's Mojave Desert, the world's largest solar power plant focuses sunlight with parabolic mirrors, vaporizing a fluid and driving generators. By 1994, this station is slated to produce 680 megawatts, two-thirds the output of a typical nuclear plant.

escape the greenhouse trap. This step is a crucial part of any worldwide plan to stem global warming. While developing nations accounted for just 7 percent of greenhouse emissions in 1950, they produce nearly 30 percent of those emissions today. By 2020, because of population growth and the ensuing rise in demand for electricity and transportation, these nations may surpass the industrialized nations in output of greenhouse gases.

Industrialized nations can help finance research into ways for the developing nations to expand their economies without unnecessarily increasing their emissions of greenhouse gases. One conduit for such funding might be the multilateral development banks, such as the World Bank, which in the past provided loans to spur economic growth, but without much regard for the environment. Another proposal has been to create a global carbon tax, which would be levied to nations in proportion to emissions of greenhouse gases.

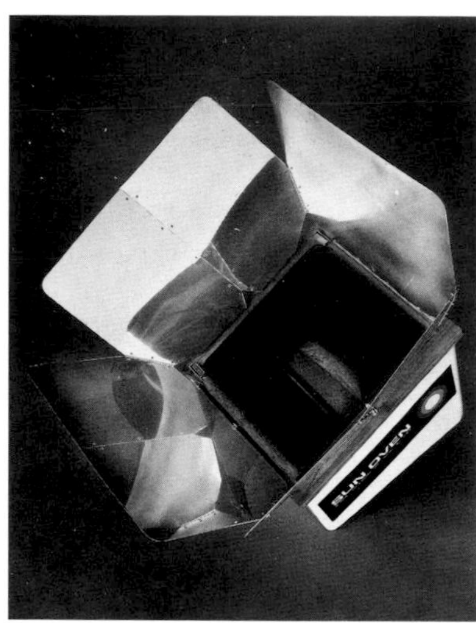

The technology for reducing greenhouse emissions need not be complicated. This solar oven, built by Burns-Milwaukee, Inc., may be a great help to people in the tropics, many of whom must now spend hours cutting firewood to cook daily meals.

Such a tax would promote cuts in fossil-fuel use and reductions in deforestation; at the same time, the revenues could finance an international environmental fund. Many roadblocks stand in the way of such a plan, one of which is the question of how to measure emissions precisely from such diverse sources as burning forests and power plants.

The World Bank and other international development institutions can also promote proven alternatives to polluting practices. For example, in the tropics, simple gas stoves or solar ovens can take the place of wood fires. In a world where more than two billion people still rely on firewood for fuel, such a simple switch could make a big difference. A United Nations study showed that the less developed nations could benefit tremendously from increased energy efficiency. A report on energy use, produced by the United Nations' World Commission on Environment and Development, said, "It is the poorest who are most often condemned to use energy and other resources least efficiently. . . . The woman who cooks in an earthen pot over an open fire uses perhaps eight times more fuel than her affluent neighbor with a gas stove and aluminum pans. The poor who light their homes with a wick dipped in a jar of kerosene get one hundredth of the illumination of a 100-watt bulb and use just as much energy to do so."

More advanced nations can also directly transfer energy-efficient technology to the developing countries and to countries in Eastern Europe where large cuts in greenhouse emissions can be made. Two such technologies are safe substitutes for CFCs and more efficient equipment for generating and distributing electric power. One example of the potential payoff: Even without replacing its existing coal-fired power plants, India could effectively double the output of the plants by replacing its antiquated system for distributing electricity.

Carbon Trading

One proposal for accomplishing such technology transfers is international "emissions trading," a concept refined by economists at the Environmental Defense Fund. The idea is to set a goal for reductions in carbon dioxide emissions worldwide as well as on a nation-by-nation basis. Each nation's cuts would be determined in proportion to, say, its economic output or population. Nations would have the option either of reducing their own emissions or helping to reduce another country's emissions. The incentive to make big cuts would be a payoff in "carbon dioxide credits," doled out to countries that cut emissions below the agreed reductions. These credits could then be bought and sold on a world market. Their value would lie in the fact that the holder would be allowed to emit some carbon dioxide.

According to Daniel Dudek, an economist with the Environmental Defense Fund, an emissions trade could work like this: Poland is extremely inefficient in its use of coal and oil because of its antiquated power plants and industries. If an American company were to sell Poland energy-efficient equipment at a sharp discount, that company could be paid in part with carbon dioxide credits—credits awarded because Poland had sharply cut its emissions. The American company could then sell the credits back in the United States to, say, an efficient electric utility—allowing it to emit a little more carbon dioxide than it might otherwise. The utility would save money because it could only achieve further cuts in carbon dioxide emissions at great expense. Overall, the large reduction in emissions in Poland would more than offset the slight extra allowance of emissions in the United States. The overall impact of such a trade would be a net reduction in the amount of carbon dioxide released into the atmosphere. As with a carbon tax, this proposal depends on negotiating an international agreement and establishing a viable system for measuring carbon dioxide emissions around the world—no easy task.

The Burden of Debt

On another front, developed nations can also reevaluate the massive foreign debt that currently burdens the less developed countries. It is unrealistic to expect poor nations to talk about efficiency and environmental conservation when they are paying off loans that were originally intended to help them develop their natural resources and economies, but now drain both their resources and their economies.

In 1988, the World Bank reported that the seventeen most indebted nations paid out $31.1 billion more in interest than they received in aid. Some debt may have to be written off altogether. A portion may be swapped for commitments to preserve endangered ecosystems. Small debt-for-nature swaps have already been accomplished in Ecuador and Central America. In 1991, Brazil offered to swap up to $100 million a year in debt. In return, Brazil would commit new funding to environmental projects.

Much of the "greenhouse effect" of poorer nations is a function of explosive population growth. Therefore, more can be done to encourage education, family planning, and public health—all of which help reduce family size. Even though the rate of growth of the human population is slowing, and

Most scientists agree that the preservation of tropical rain forests should be a worldwide priority. The destruction of these forests adds to the buildup of greenhouse gases: The burning and rotting vegetation emits carbon dioxide and methane. The rain forests are also crucial reservoirs of biological diversity, harboring a vast array of species including this collection of mammals (clockwise from top left): the golden anteater of Venezuela; Goodfellow's tree kangaroo, from Papua New Guinea; the squirrel monkey of Central America; and the capybara of Venezuela.

should stabilize late in the coming century, this trend is too slow to be of any comfort to those monitoring the steady rise of carbon dioxide, methane, and the other heat-trapping gases. In Africa, the median age is only seventeen, and the size of the typical family—with six or more children—has not diminished substantially in four decades.

Deforestation and Reforestation

Population pressure has greatly increased the amount of burning of vegetation to clear land for pasture or fields. Inequitable distribution of land in many countries adds to the problem. In Brazil, millions of poor people have been pushed from fertile land in the developed south as a wealthy elite accumulates vast holdings; the poor have thus migrated into the Amazon wilderness. The result of all of such pressures, according to NASA scientist Joel Levine, is that somewhere between 2 and 5 percent of the planet's land area burns every year. Satellites annually record a broad glimmering band of fires ringing the globe near the equator, through Africa, Asia, and South America.

Along with eliminating some of the causes of deforestation, which emits torrents of carbon dioxide and methane, every nation could begin reforestation programs. As photosynthesis takes place in growing trees, carbon dioxide is broken down into carbon and oxygen, and the carbon is locked away in the tissue of the tree for decades, if not centuries. Recently, an American electric

Along with preventing deforestation, some communities are initiating reforestation projects. In the Brazilian Amazon, workers tend to tree seedlings destined for denuded tracts.

Reforestation makes sense for many reasons. In French Polynesia, the government is planting trees to prevent erosion and provide a renewable supply of timber. A worker carries pine seedlings along a ridgetop on the island of Raiatea (right).

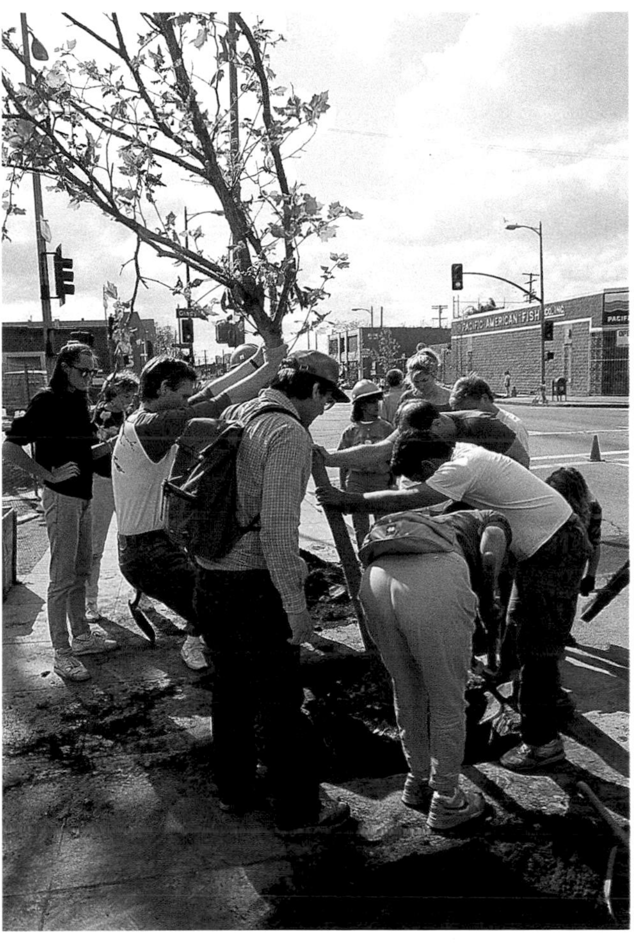

Many private groups are promoting tree planting in the United States. The Los Angeles organization TreePeople adds a patch of shade to the city (left). Urban trees help reduce pollution and can cut air-conditioning costs.

utility, Applied Energy Services, conducted something of a voluntary emissions trade involving trees. To offset anticipated carbon dioxide emissions from a new power plant in Connecticut, the company paid $2 million to plant 52 million trees in Guatemala. At a low cost, the trees will pull more carbon dioxide from the air than the power plant is projected to emit over its forty-year lifetime.

There are secondary benefits to tree planting. Trees have a cooling effect on their immediate surroundings and can significantly reduce the demand for air conditioning when planted near houses or buildings. And of course forests hold moisture and soil, preventing further degradation of the landscape.

California's Global Agenda

This list of actions to soften the greenhouse impact appears daunting, but there is ample evidence from around the world that most of these proposals can work. The United States government has been slow to adopt measures to reduce carbon dioxide emissions, but the federal bureaucracy is not the only entity with the power to change the national agenda. Decisions made at the state level can have national repercussions, particularly when the state is California. In 1990, to foster a shift to more efficient automobiles, the state legislature overwhelmingly passed a law that would create a system called Drive Plus. The idea? As the owners of new cars register their vehicles, they are either assessed a fee or given a rebate, depending on both the car's emissions and fuel efficiency. Because the fees paid by owners of gas guzzlers pay for the rebates offered to owners of efficient cars, the program finances itself—requiring no new taxes. The bill was vetoed by California's outgoing governor the first time around, but predictions were that a plan like it could take hold in the near future.

One recent California law that did pass the first time around mandates that by 1998, all auto manufacturers selling their product in the state must have at least 2 percent of their fleets operate with *no tailpipe emissions* at all. The only cars now in existence that can satisfy this requirement are, of course, electric cars. Thanks largely to the California statute—and the assumption that federal standards will soon be more stringent—General Motors announced in March 1991 that it was converting a small plant in Lansing, Michigan, to the production of the first commercial electric car since the early years of this century.

There have been dramatic regional examples showing the benefits of planting trees and demonstrating that it is possible to reduce deforestation. In

In Brazil and other nations with territory in the Amazon river basin, local communities are showing that some of the rain forest's bounty can be harvested without cutting down the trees. Below, an Indian in the Brazilian state of Rondônia slashes the bark of a rubber tree to collect latex. The tree is not harmed and will produce latex for many years.

one Los Angeles neighborhood, trees were planted and the houses were also painted light colors and roofed with light-colored material; as a result, the use of air conditioners dropped almost 50 percent. This experiment demonstrates how tree planting in cities may help prevent yet another unwelcome greenhouse feedback loop, in which more warming creates higher demands for energy to keep homes and offices cool; and these demands lead to more emissions and more warming.

Fighting the Fire

In 1987, a smoke cloud the size of India hung over the Brazilian part of the Amazon river basin. Since then, increased forestry patrols and the elimination of some tax incentives that encouraged destructive development have cut the rate of deforestation in Brazil by more than half from that devastating peak.

The Rocky Mountain Institute in Colorado is not only a center for the study of energy efficiency; it is also an example of energy-efficient architecture. The stone buildings, with super-insulating glass windows, have extremely low heating costs.

The Brazilian government has set aside tracts of forest—a total area the size of Massachusetts—as "extractive reserves." In these reserves, deforestation is prohibited and local communities can safely harvest valuable forest products such as rubber and nuts. In a nation that lost an area of forest twice the size of California in just a decade, one Massachusetts-size reserve is a small step, but it is a step in the right direction.

New Standards of Wealth

The challenge posed by global warming also presents an opportunity for reconsidering the means by which progress is measured in contemporary industrial societies. Traditionally, the health of an economy has been judged in large part by the annual growth of the gross national product. In such calculations, the loss column has never taken into account the cost of degradation

Some scientists say that humans must begin to take into account their impact on other species not just for utilitarian reasons, but also for ethical reasons. This chamois thrives in the coldest, uppermost reaches of the French Alps. Global warming may shift local ecosystems upward to the extent that the species, forced out of its range, has nowhere to go.

of natural resources or ecosystems or damage to the atmosphere or public health.

No one has more aptly described this problem than José Lutzenberger, Brazil's secretary for the environment. In 1990, Secretary Lutzenberger addressed a meeting of more than a hundred lawmakers from around the world who convened in Washington to discuss ways to halt the degradation of the atmosphere, oceans, and water supplies. As Lutzenberger put it, some people seem to think that "the economy is a flow in a single direction between two infinities—infinite resources on one side, and an infinite hole on the other side into which we can dump all our wastes."

As an example of the way the world currently works, Lutzenberger described Brazil's crash program to develop the vast mineral deposits deep in the Amazon rain forest. "When we in Brazil export iron ore and aluminum, we add up the foreign exchange that the export brings us, but we do not subtract the loss of ore, the demolition of the mountain, of the forest, the genocide of the Indians and all the other losses." If these losses were added up, the destruction of the forested mountain suddenly would not make nearly as much sense.

He concluded, "We forget that economics is only a chapter of ecology. Economics deals with the interactions and the flow of resources between humans. Ecology deals with life as a whole, of which we humans are only a part. So, good, true economics must be based on good ecology."

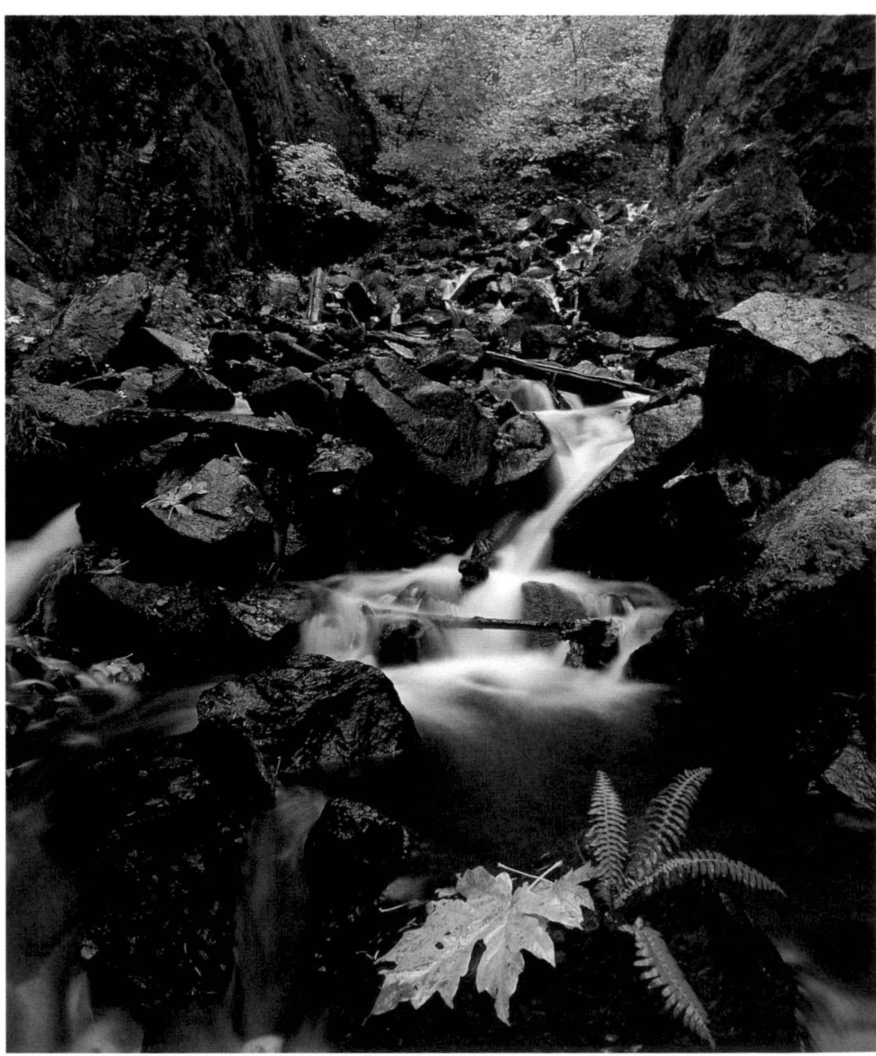

Perhaps the simplest step a person can take to begin to counter the greenhouse effect is to reestablish a connection with the natural world—to start with the basic act of taking a walk in the woods. The rewards are evident in this view of the Columbia River Gorge National Scenic Area, in Oregon.

Walking Softly on the Earth

Finding a way to change the system governments use to measure economic progress, or to accomplish any of the many changes needed to reduce the impact of global warming, will not be easy. Indeed, some politicians claim that there is simply no way to convince a large part of humanity to change deeply ingrained habits in order to offset a disruptive, human-caused climate change one, two, even three generations into the future.

There are people at every level of society, however, and in societies around the world, who are already demonstrating that it is possible for the human species to flourish and, at the same time, walk a little more softly on the Earth. These people include the California legislators who found a way to push auto manufacturers to start building electric cars. They include the rubber-tree tappers and Indians of the Amazon rain forest who are showing their nation's planners how it may be possible to reap a harvest from the forest without harming the trees. They are scientists such as Charles David Keeling, who has devoted his career to charting the inexorable rise of atmospheric carbon dioxide levels—for the first time providing tangible evidence that humans are fiddling with the planetary thermostat. They are politicians such as Gro Harlem Brundtland, the prime minister of Norway, who has become an international ambassador for altering the formula for economic development so that it includes environmental preservation.

These people are also the American inventors and entrepreneurs who continued to tinker with solar cells, alternative fuels, windmills, and other energy-saving technologies through the days when budget cuts ended most public funding for such work. They are homeowners who took out a calculator and discovered that it might just make sense to spend $13 for a compact fluorescent light bulb.

They are people who are taking the most basic actions to soften the human impact: planting a tree on their city block; riding a bicycle to work on a sunny day; turning off the lights when they leave a room; or simply taking a long walk in the woods with their sons and daughters, to remind themselves what this is all about.

"Because it's a new category, they haven't settled the details of what to do with those who have sinned against the planet."

Drawing by Donald Reilly.
© 1990 The New Yorker Magazine, Inc.

Abrahamson, Dean Edwin, ed. *The Challenge of Global Warming.* Washington, D.C.: Natural Resources Defense Council and Island Press, 1989.
Scientists and policy experts offer explanations of the causes and potential impacts of global warming and recommend actions to counter it.

Collins, Mark, ed. *The Last Rain Forests.* New York: Oxford University Press, 1990.
A lushly illustrated atlas of Earth's few remaining reservoirs of biological riches.

Clark, Sarah L. *Protecting the Ozone Layer: What You Can Do.* New York: Environmental Defense Fund, 1988.
A brief but informative look at how CFCs destroy ozone, what these chemicals are used for, and what individuals and governments can do to reduce their use.

Committee on Science, Engineering, and Public Policy of the National Academy of Sciences, National Academy of Engineering, and Institute of Medicine. *Policy Implications of Greenhouse Warming.* Washington, D.C.: National Academy Press, 1991.
A sober analysis of the global warming problem by a balanced panel of experts, and a recommendation that some actions be taken now to mitigate adverse effects.

Ehrlich, Paul and Anne. *The Population Explosion.* New York: Touchstone Books, 1991.
A troubling account of the continuing growth in the human population and its impact on the planet.

Firor, John. *The Changing Atmosphere: A Global Challenge.* New Haven: Yale University Press, 1990.
A slim, rich volume describing how humans have altered the atmosphere.

Houghton, J.T., G.J. Jenkins, and J.J. Ephraums, eds. *Climate Change: The IPCC Scientific Assessment.* New York: Cambridge University Press, 1990.
The most comprehensive analysis of the greenhouse effect yet published.

Leggett, Jeremy, ed. *Global Warming: The Greenpeace Report.* New York: Oxford University Press, 1990.
A collection of essays by leading scientists and energy analysts on the science, impacts, and policy issues related to global warming.

Lyman, Francesca, ed. *The Greenhouse Trap.* Boston: Beacon Press, 1990.

A concise explanation of what is known and what remains uncertain about the greenhouse effect and what governments and individuals can do to mitigate global warming.

McKibben, Bill. *The End of Nature.* New York: Anchor Books, 1990.
An elegant essay on the pervasive impact of human beings on the atmosphere, the Earth, and the biosphere.

Oppenheimer, Michael, and Robert H. Boyle. *Dead Heat: The Race Against the Greenhouse Effect.* New York: Basic Books, 1990.
A fluently written account of the greenhouse problem and a persuasive prescription for action.

Revkin, Andrew. *The Burning Season: The Murder of Chico Mendes and the Fight for the Amazon Rain Forest.* Boston: Houghton Mifflin, 1990.
A chronicle of events leading to the assault on the Amazon and the murder of one of the forest's most effective defenders.

Roan, Sharon. *The Ozone Crisis: The Fifteen Year Evolution of a Sudden Global Emergency.* New York: John Wiley & Sons, 1990.
An excellent description of the research that uncovered the depletion of the stratospheric ozone layer, and of the disturbing implications of this trend.

Schneider, Stephen H. *Global Warming: Are We Entering the Greenhouse Century?* San Francisco: Sierra Club Books, 1989.
A clear, informative appraisal of the scientific evidence for global warming, written by a longtime student of links between climate and human affairs.

Weiner, Jonathan. *The Next One Hundred Years: Shaping the Fate of Our Living Earth.* New York: Bantam Books, 1990.
An enlightening account of current environmental issues, particularly global warming and ozone depletion, and of how generations of scientists have contributed to our understanding of these problems.

World Resources Institute, in collaboration with the United Nations Environment Programme. *World Resources 1990–91.* New York: Oxford University Press, 1990.
An exhaustive compendium of data on such factors as population growth, greenhouse-gas emissions, the global water supply, and deforestation.

Worldwatch Institute. *State of the World 1991.* New York: W. W. Norton & Co., 1991.
The most recent in a series of detailed annual reports assessing progress toward a sustainable society.

A GREENHOUSE DIET

20 Things You Can Do to Reduce the Risk of Global Warming

Here are some simple steps you can take starting today to become part of the solution to the greenhouse problem.

Whenever you save energy—or use it more efficiently—you not only save money, you also reduce the demand for such fossil fuels as coal, oil, and natural gas. Less burning of fossil fuels means lower emissions of carbon dioxide, the primary contributor to global warming.

The average American weighs in at about 40,000 pounds of carbon dioxide (CO_2) emissions per year. By exercising even a few of these steps, you can cut your annual emissions by thousands of pounds!

Home Appliances

1 Make sure your dishwasher is full when you run it and use the energy-saving setting, if available, to allow the dishes to air dry. You can also turn off the drying cycle manually. Not using heat in the drying cycle can save 20 percent of your dishwasher's total electricity use. This simple step reduces your CO_2 emissions by 200 pounds per year.

2 Wash clothes in warm or cold water, not hot. Switching from hot to warm for two loads per week can save nearly 500 pounds of CO_2 per year if you have an electric water heater, or 150 pounds per year for a gas heater.

3 Turn down your water heater thermostat. Thermostats are often set to 140°F when 120°F is usually fine. Each 10-degree reduction saves 600 pounds of CO_2 per year for an electric water heater, or 440 pounds per year for gas.

Home Heating and Cooling

4 Be careful not to overheat or overcool rooms. Lowering your thermostat just two degrees during winter saves 6 percent of heating-related CO_2 emissions. That's a reduction of 420 pounds of CO_2 per year for a typical home.

5 Clean or replace air filters as recommended. Energy is lost when air conditioners and hot-air furnaces have to work harder to draw air through dirty filters. Cleaning a dirty air conditioner filter can save 5 percent of the energy used. That could save 175 pounds of CO_2 per year.

Small Investments That Pay Off

6 Buy energy-efficient compact fluorescent bulbs for your most-used lights. Although they cost more initially ($15 to $20 apiece), these bulbs save money in the long run by using only one-quarter the energy of an ordinary incandescent bulb to produce the same amount of light. In a typical home, one compact fluorescent bulb can save 260 pounds of CO_2 per year.

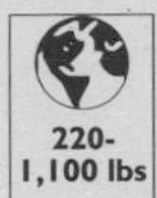

7 Wrap your water heater in an insulating jacket, which costs just $10 to $20. It can save 1,100 pounds of CO_2 per year for an electric water heater, or 220 pounds per year for gas.

8 Use less hot water by installing low-flow shower heads. They cost just $10 to $20 each, deliver an invigorating shower, and save 300 pounds of CO_2 per year for electrically heated water, or 80 pounds per year for gas.

9 Weatherize your home or apartment, using caulk and weather stripping to plug air leaks around doors and windows. Caulking costs less than $1 per window, and weather stripping costs under $10 per door. These steps can save up to 1,100 pounds of CO_2 per year for a typical home.

10 Ask your utility company for a home energy audit to find out where your home is poorly insulated or energy inefficient. This service may be provided free or at low cost. Make sure it includes a check of your furnace and air conditioning.

Getting Around

11 Whenever possible, walk, bike, carpool, or use mass transit. Every gallon of gasoline you save avoids 22 pounds of CO_2 emissions. If your car gets 25 miles per gallon, for example, and you reduce your annual driving from 12,000 to 10,000 miles, you'll save 1,800 pounds of CO_2.

12 When you next buy a car, choose one that gets good mileage. If your new car gets 40 miles per gallon instead of 25, and you drive 10,000 miles per year, you'll reduce your annual CO_2 emissions by 3,300 pounds.

Reduce, Reuse, Recycle

13 Reduce the amount of waste you produce by buying minimally packaged goods, choosing reusable products over disposable ones, and recycling. For every pound of waste you eliminate or recycle, you reduce emissions of CO_2 by at least 1 pound. Cutting down your garbage by half of one large trash bag per week saves at least 1,000 pounds of CO_2 per year.

14 If your car has an air conditioner, make sure its coolant is recovered and recycled whenever you have it serviced. In the United States, leakage from auto air conditioners is the largest single source of emissions of chlorofluorocarbons (CFCs), which damage the ozone layer as well as add to global warming. The CFCs from one auto air conditioner can add the equivalent of 4,800 pounds of CO_2 emissions per year.

Home Improvements

When you plan major home improvements, consider some of these energy-saving investments. They save money in the long run, and their CO_2 savings can often be measured in tons per year.

15 Insulate your walls and ceilings. This can save 20 to 30 percent of home heating bills and reduce CO_2 emissions by 1,400 to 2,100 pounds per year. If you live in a colder climate, consider superinsulating. That can save 5.5 tons of CO_2 per year for gas-heated homes, 8.8 tons per year for oil heat, or 23 tons per year for electric heat. (If you have electric heat, you might also consider switching to more efficient gas or oil.)

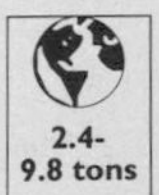

16 Modernize your windows. Replacing all your ordinary windows with argon-filled, double-glazed windows saves 2.4 tons of CO_2 per year for homes with gas heat, 3.9 tons for oil heat, and 9.8 tons for electric heat.

17 Plant shade trees and paint your house a light color if you live in a warm climate, or a dark color if you live in a cold climate. Reductions in energy use resulting from shade trees and appropriate painting can save up to 2.4 tons of CO_2 emissions per year. (Each tree also directly absorbs about 25 pounds of CO_2 from the air annually.)

18 As you replace home appliances, select the most energy-efficient models. Replacing a typical 1973 refrigerator with a new energy-efficient model, for example, saves 1.4 tons of CO_2 per year. Investing in a solar water heater can save 4.9 tons of CO_2 annually.

Business and Community

19 Work with your employer to implement these and other energy-efficiency and waste-reduction measures in your office or workplace. Form or join local citizens groups and work with local government officials to see that these measures are taken in schools and public buildings.

20 Keep track of the environmental voting records of candidates for office. Stay abreast of environmental issues on both the local and national levels, and write or call your elected officials to express your concerns about energy efficiency and global warming.

For More Information

Clark, Sarah L., **Fight Global Warming: 29 Things You Can Do** (New York: Consumer Reports Books in association with Environmental Defense Fund, 1991).

DeCicco, John, et al, **CO_2 Diet for a Greenhouse Planet: A Citizen's Guide for Slowing Global Warming** (New York: National Audubon Society, 1990).

Earthworks Group, **30 Simple Energy Things You Can Do to Save the Earth** (Berkeley, Calif.: Earthworks Press, 1990).

Wilson, Alex, **1991 Consumer Guide to Home Energy Savings** (Washington, DC: American Council for an Energy-Efficient Economy, 1990).

"Laugh if you will, but my kind once ruled the earth."

Drawing by Jack Ziegler
© 1991 The New Yorker Magazine, Inc.

Alliance to Save Energy
1725 K St. NW, Washington, DC 20006; 202-857-0666
Dedicated to increasing energy efficiency, this group conducts research and pilot projects to evaluate solutions to energy-use problems.

American Council for an Energy Efficient Economy
1001 Connecticut Ave. NW, Washington, DC 20036; 202-429-8873
Publishes *The Most Energy Efficient Appliances,* an annual guide, and other guides to home energy savings.

American Forestry Association
P.O. Box 2000, Washington, DC 20013; 202-667-3300
A national citizens organization dedicated to the conservation, maintenance, and improvement of forests. Sponsors "Global Re-Leaf," a national program that encourages Americans to plant millions of trees.

Environmental Defense Fund
257 Park Avenue South, New York, NY 10010; 212-505-2100
This group's scientists, economists, and lawyers work to defend the environment by fostering informed congressional and public debate; providing credible scientific analysis; and devising innovative, economically viable solutions to environmental problems.

Friends of the Earth
530 7th St. SE, Washington, DC 20003; 202-544-2600
Promotes the conservation, protection, and rational use of the Earth's resources. Its activities include lobbying, litigation, and public information on a variety of environmental issues, including ozone depletion, river protection, and tropical deforestation.

Greenhouse Crisis Foundation
1130 17th St. NW, Suite 630, Washington, DC 20036; 202-466-2823
A project of the Foundation on Economic Trends, this group is dedicated to creating global awareness of the greenhouse crisis and to changing the world view and life-styles underlying that problem.

Greenpeace USA
1436 U St. NW, Washington, DC 20009; 202-462-1177
An action organization dedicated to preserving the Earth and the life it supports,

halting the needless slaughter of marine mammals and other endangered species, protecting against nuclear and toxic pollution, and stopping the threat of nuclear war.

League of Conservation Voters
320 4th St. NE, Washington, DC 20002; 202-785-8683
A national, nonpartisan political arm of the environmental movement. It works to elect pro-environmental candidates to Congress, based on energy, environment, and natural resource issues. Publishes *The National Environmental Scorecard*, an annual rating of members of Congress on environmental issues.

National Arbor Day Foundation
100 Arbor Ave., Nebraska City, NE 68410; 402-474-5655
This organization promotes tree planting and conservation by providing direction, technical assistance, and public recognition for urban and community forestry programs.

National Audubon Society
950 Third Ave., New York, NY 10022; 212-832-3200
Through research, education, and action, this group works for the long-term protection and wise use of land, water, and other natural resources.

National Wildlife Federation
1400 16th St. NW, Washington, DC 20036-2266; 202-797-6800
A nonprofit conservation group dedicated to encouraging global awareness of the need for informed use and proper management of natural resources and wildlife. Carries out a strong educational program.

Natural Resources Defense Council
40 West 20th St., New York, NY 10011; 212-727-2700
Aims to protect America's endangered natural resources and to improve the quality of the human environment. It combines litigation and science in monitoring government agencies, bringing legal action, and disseminating information.

The Nature Conservancy
1815 N. Lynn St., Arlington, VA 22209; 703-841-5300
Acts to preserve ecosystems and the rare species and communities they shelter. The group has protected more than 3.5 million acres of threatened habitat, mostly by purchasing land, and manages more than one thousand preserves.

Rainforest Action Network
300 Broadway, Suite 28, San Francisco, CA 94133; 415-398-2732
This group focuses exclusively on rain forest protection, working with other

environmental and human rights organizations on major rain forest education and protection campaigns.

Rainforest Alliance
270 Lafayette St., Suite 512, New York, NY 10012; 212-941-1900
A membership organization committed solely to the conservation of the world's tropical forests. It seeks positive solutions to deforestation through public education, research, the development of private sector initiatives, and direct support to peoples living in the forest.

Resources for the Future
1616 P St. NW, Washington, DC 20036; 202-328-5000
Conducts research on the environment and the conservation and development of natural resources. Issues it addresses include air and water pollution, solid waste disposal, pesticides, toxic substances, and international issues.

Rocky Mountain Institute
1739 Snowmass Creek Rd., Old Snowmass, CO 81654; 303-927-3128
Aims to foster the efficient and sustainable use of resources as a path to global security. It offers its research on resource efficiency, global security, and community economic revival through its publications. A publication list is available upon request.

Sierra Club
730 Polk St., San Francisco, CA 94109; 415-776-2211
Founded in 1892 to explore, enjoy, and protect the wild places on Earth; to practice and promote the responsible use of the Earth's ecosystems and resources; and to educate and enlist humanity to protect and restore the quality of the natural and human environment.

Union of Concerned Scientists
26 Church St., Cambridge, MA 02238; 617-547-5552
An organization of scientists and other citizens concerned about the impact of advanced technology on society. Its efforts focus on national energy policy, nuclear safety, nuclear arms control, and global warming.

The Wilderness Society
900 17th St. NW, Washington, DC 20006-2596; 202-833-2300
A nonprofit membership organization devoted to preserving wilderness and wildlife; protecting America's prime forests, parks, rivers, and shorelands; and fostering an American land ethic.

World Resources Institute
1735 New York Ave. NW, Washington, DC 20006; 202-638-6300

Helps government, the private sector, and organizations address issues in environmental integrity, resource management, economic growth, and international security. Each year it publishes the *World Resources Report*.

Worldwatch Institute
1776 Massachusetts Ave. NW, Washington, DC 20036; 202-452-1999
A nonprofit research organization that works to identify emerging global problems and trends and bring them to the attention of opinion leaders and the general public.

World Wildlife Fund
1250 24th St. NW, Washington, DC 20037; 202-293-4800
Works to protect endangered wildlife and wildlands. Its top priority is conservation of the tropical forests in Latin America, Asia, and Africa.

INDEX

(Page numbers in *italic* refer to illustrations.)

agriculture, 18, *64*, 124
air conditioning, *99*, 146–47, *159*, 160, 161
air pollution, *19–22*, 60–61
Alexander VI, Pope, 50
algae, 42, *42*, *43*
Alps, 52, *52*, *53*, *163*
aluminum, 163–64; recycling of, *146–47*, 147
Amazon rain forest, 39, *76*, *80*, 143; conservation efforts in, *158*, 161–62, *161*, 165; destruction of, *2–3*, *19*, 23, 77–82, *78*, 158; mineral deposits in, 163–64
ammonia, 41
Antarctica, 46, 69–70, *68–70*, 77, *100–101*, 103, 112, 143; hole in ozone layer over, 113–14, *113*
Applied Energy Services, 160
architecture, energy-efficient, *142*, *162*
Arctic, 24, *102*, *111*, *114*, 121, 143
Arctic National Wildlife Refuge, 120
Arrhenius, Svante, 57, 72–73, 77, 86, 91
atmosphere: in constant flux, 33; heat-trapping properties of gases in, 62–63; history of, 41–52; overall stability and predictability of, 33; tampering with, 71–73; viewed from satellites and space craft, *27*, 33, *38*, *39*; workings of, 23–29
automobiles, 17, *64*, 77, 82, *85*, 88, 90–91, 121, *122*, 148–50, 160, 165; electric, *150*, *151*, 160; gas-mileage standards for, 148–49

bacteria, 42, 95
Ball, Ben C., Jr., 149
Bangladesh, 130, *131*
bicycles, *149*
blue-green algae, *42*, *43*
Bonneville Salt Flats, 39
Botkin, Daniel, 139
Brazil, 155; conservation efforts in, 158, 161–62, *161*; destruction of rain forest in, *2–3*, *19*, 23, 77–82, *78*, 158; mineral deposits in, 163–64
Broecker, Wallace, 112–13
Brooklyn Botanic Garden, 61–62, *62*
Brundtland, Gro Harlem, 165
Buddemeier, Robert, 130
Burns, Robert, 75
Burns-Milwaukee, Inc., *154*
"business-as-usual" scenario, 120–23, *124*

California, 160–61, 165. *See also* Los Angeles
Callendar, George, 91, 94
carbon, 46
carbon dioxide, 16, 18, 26, 41, 45–46, 59, *71*, *95*, 97, 99, 103, 143, 158; agriculture and, 124; computer models of effects of, 103–7, *104*, 108–10, 112; greenhouse effect and, 61, 62, 63, *64–65*, 66, 67; from human activities, 17, 55, 71–73, *72*, 78, 88, *88*, *89*, 90, 144; measuring concentrations of, 92–93, *92*, 105; in photosynthesis, 42, 124; reducing emissions of, 149–50, 151, 154–55, 158, 160; rise in levels of, 93–94, *94*, 165; temperature and, 69–71, *70*, 73, 91, 94–95
Carboniferous Period, *45*, 70–71
carbon tax, 153–54, 155
carbon trading, 154–55
cattle, *64*, 95, *97*
Celebrations of Life (book; Dubos), 90–91
charcoal, 73, 87
China, 49, 78, 120–21, *121*
chlorofluorocarbons (CFCs), 17–18, *64–65*, *95*, 143; cuts in production of, 144, 154; in degradation of ozone layer, 77, 99, 113, *113*, 114–15, 144; sources of, *64–65*, 97–99, *99*
cirrus clouds, *109*
cities, "heat-island" effect of, 94
CleanAir, *151*
climate: changes in, 15–16, 36, 50–52; clouds' effects on, 107–8, *109*; computer models of, 103–13, *104*, 121, 123, *123*, 131; cycles in, 46–49, *49*; effects of human activities on, 16–18, 55, *64–65*, 67, 71–73, *72*; evidence of dramatic change in, 38–39, *40*, *41*; predictable patterns in, 33–36
clouds, 33, 104, 110, 112; climate affected by, 107–8, *109*
coal, 17, 22, 42, 70, 71, 73, 86, 120–21, 150, 155
coal mining, *72*, *86*, *87*, 95
coastal encroachment, 18, *128*, 130–33, *131–33*
computer models, 18, 103–13, *104*, 121, 123, *123*, 131; clouds and, 108; evaluating of, 107, 108–10; sea level and, 110–12; unpredictable responses of nature and, 112–13; working of, 104–7
cooking fires, 88, *89*, 154; solar ovens and, *154*
coral, 22, 49, *83*
Cowley, Abraham, 31
crocodiles, 39, *41*
cumulus clouds, 108, *109*

debt-for-nature swaps, 155
deforestation, 65, 130, 164; in rain forests, 17, *19*, 23, *74–75*, 77–82, *78–79*, 121, 158; reducing of, 154, *158*, 160–62, *161*
developing nations, 22, 120, 152–54; debt-for-nature swaps with, 155; population growth in, 155–58
dinosaurs, 46, *47*, *54*, 55, 139
Drake, E.L., *87*
Drive Plus, 160
droughts, 18, *123*, 125, *126–27*
Dubos, René, 90–91
Dudek, Daniel, 155
dust devils, 33, *33*

Earth: axis of, 105; influence of life forms on, 42–44; orbit and rotation of, 36, 48; scarred by human activities, *74–75*, 76–86, *78–86*
Eastern Europe, 21, 22, 154, 155
economic progress, 22, 162–63, 165

INDEX

Ecuador, 155
Egypt, 125, 130
Eiseley, Loren, 87, 141
electric cars, *150, 151,* 160
electricity, 145–47, *145,* 153, 154; nuclear power and, 150, 151–52; solar or wind power and, *140–42,* 150–51, *152, 153*
elephants, African, *25*
Emanuel, Kerry, 133
emissions trading, 154–55
energy efficiency, 145–52; in American factories, 145–46; insulation and, 146; light bulbs and, 146–47, 165; nuclear power and, 150, 151–52; recycling and, *146–47,* 147; solar or wind power and, *140–42,* 150–51, *152, 153;* in transportation, 148–50, *149, 151*
Environmental Defense Fund, 154–55
Environmental Protection Agency, 131
Everglades National Park, *137*
Exeter (R.I.), *14,* 15
extinctions, 139; mass, 46–48, 55
factories, *21,* 145–46
fern forests, *45*
fertilizers, 97, *98*
fires, 49, *65, 90,* 158; for cooking, 88, *89,* 154; forest (summer of 1988), *58,* 59, *59;* at Kuwaiti oil wells, *116–17, 118,* 119–20; in rain forests, *2–3,* 17, *19,* 23, *74–75,* 77–82, *78–79,* 121, 158
Fishing in American Waters (book; Scott), 77
fluorescent light bulbs, compact, 147, 165
foreign debt, 155
forests, 133–36, *135,* 139; fires in, *58,* 59, *59, 65. See also* rain forests
Fourier, Jean-Baptiste-Joseph, 62, 72
France, *21,* 50, 52, *52, 53,* 59, 151
Frost Fairs, 52, *53*

Galveston (Tex.), 131, *132*
gas-mileage standards, 148–49
gasoline, 120; lead additive for, 97, 98; taxes on, 148, 149
General Motors, 160
Geophysical Fluid Dynamics Laboratory, 110, 124
Germany, 151
glaciers, *14,* 15, 18, 39, 46, 52, *52, 53, 100–102,* 103, 110, 135
global warming: adaptation to, 55; "business-as-usual" scenario and, 120–23, *124;* coastal encroachment and, 18, *128,* 130–33, *131–33;* countermeasures against, 22–23, 55, 143–65; cynicism about, 36; growing acceptance of theory of, 59–60; heat waves and, 123–24, *125;* as invisible problem, 23, 60–61; islands at risk in, 128–30, *129;* precipitation patterns and, 110, 112, 121, *123,* 124–25, *126–27;* predictions about, 18–22, 103–15, *104,* 120–39, *123, 124;* previous climatic changes vs., 16; reluctance to take action against, 143; sea level and, 18, 110–13, 121–23; "signals" of, 18; wilderness and wildlife affected by, 133–39, *135–39*
Goddard Institute for Space Studies, 94, 103, 104, 107, 124
Goldilocks phenomenon, 69
Grand Tetons, *37*
grasslands, burning of, *65*
Green Decade, 119
greenhouse effect, 16, 59–60, 61–69, 72; heightened concern about, 94–95; overall description of, 61–63; physics of, 63–69, *64–65;* planetary comparisons and, 67–69; relative contribution of gases to, *95*
Greenland, 50, *50, 51*
Gulf Stream, *105,* 107

Hansen, James, 59–60
Hawke, Bob, 130
"heat-island" effect, 94
heat waves, 123–24, *125*
Holgate Glacier, *102*
Holocene Epoch, 49–50
homes, energy efficiency in, *142,* 146–47
hurricanes, 18, 33, 131–33, *132, 134*

ice, 110, 112. *See also* glaciers
ice ages, 39, 48–49, *49,* 70, 135
icebergs, 27, *102,* 103, 110, *111*
Iceland, 50, *50*
impalas, *25*
incandescent light bulbs, 146–47
Indonesia, destruction of rain forests in, *74–75, 79*
Industrial Revolution, 17, 70, 71
infrared radiation, 63, *64–65,* 66, 104
insulation, 146
interglacials, 48–49, *49,* 70
Intergovernmental Panel on Climate Change, 108–9
International Geophysical Year, 92–93
Iron Age, 87
islands, 128–30, *129*

Japan, environmental technologies in, 148, 151, *152*
jet streams, 33, *39*
Keeling, Charles David, 92–93, 94, 165
Keeling Curve, 93, *94,* 95, 99
Kiribati, 130
Kirtland's warbler, 139, *139,* 143
Kuwait, oil well fires in, *116–17, 118,* 119–20

land distribution, 158
landfills, *64,* 95
lead, 97, 98
Leahy, Patrick, 117
Leatherman, Stephen, 130–31
Levine, Joel, 158
life: appearance of, 41–42; Earth influenced by, 42–44; and move from sea to land, 45
lighting, 146–47, 154, 165
Little Ice Age, 50–52, 71
livestock, *64,* 95, *97*
Los Angeles (Calif.), 84–86, *84,* 151, *159,* 161
Lovins, Amory, 146
Lutzenberger, José, 23, 163–64
Luz International, 150

McElroy, Michael, 23
McKibben, Bill, 115
Maldives, 128–30, *129*
mammals, 46, *47, 156–57*
manatees, *137*
Manikfan, Hussein, 128–30
manometers, 93
"Man the Firemaker" (essay; Eiseley), 87
Mars, 16, 67–69, *67*
mass transit, 148, 149, *149*
Maury, Matthew Fontaine, 23
Medieval Optimum, 50
meteorites, 46, *48*
methane, 17, 18, 41, 63, *64–65,* 66, 95, 99, 144, 158; sources of, 95–97, *96–97*
Miami (Fla.), 131, *133*
Midgley, Thomas, Jr., 97–98, 99
Montreal Protocol (1987), 144
mountain ranges, 36, *37*

Narragansett Bay, 15, 16, *16*
NASA, 103, *104*
National Academy of Sciences, 144
National Center for Atmospheric Research, 107
National Energy Strategy, 120
national parks, 139
natural gas, 17, 71, 82, 150. *See also* methane
Nile Delta, 130
Nile River, *39, 41*, 125
El Niño, 107
nitrogen, 45–46, 62–63
nitrous oxide, 17, *64–65, 95*, 99; sources of, 97, *98*
North Pole, 110, 114
nuclear power, 150, 151–52

oases, 39
ocean currents, 36, 46, 104, *105, 106*, 107, 112
oceans, 18, 41, 42, *44*, 91, *105, 106*, 107, 110, 112, 143
oil, 17, 22, 42, 71, 121, 150, 155; Kuwaiti fires and, *116–17, 118*, 119–20; production of, *20*, 82, *82, 87*, 95, 120; U.S. consumption of, 17, 145, 146, 148–49
oil spills, 119
Oppenheimer, Michael, 115
Ordovician Period, *44*
oxygen, 26, 42–44, 45–46, 55, 62–63
ozone layer, 18; creation of, 44–46; degradation of, 77, 99, 113–15, *113*, 144

panther, Florida, *138*, 139
penguins, Adelie, *106*
Persian Gulf War, *116–17, 118*, 119–20
photosynthesis, 42–44, 55, 93, 124
Physical Geography of the Sea, The (book; Maury), 23–29, *26*
plankton, 143
Pleistocene, *47*, 48–49, 143
Poland, 155
poles, 46, 113, 114, 124; ice caps at, 39, 48, 104, 110, 112
pollution: air, *19–22*, 60–61; water, 60
Pope, Alexander, 101
population growth, 17, 55, 86–87, *88*, 94, 95, 121, 153, 155–58
power plants, *65*, 77
precipitation, 26–27, 33, *38*, 41, 130; global warming and, 110, 112, 121, *123*, 124–25, *126–27*
Preparing for Climate Change (book; Topping and Helm), 22

railways, 149
rain. *See* precipitation
rain forests, 39, *76*, 143; Brooklyn Botanic Garden exhibit of, 61–62, *62;* conservation efforts in, *158*, 161–62, *161*, 165; destruction of, 17, *19*, 23, *74–75*, 77–82, *78–79*, 121, 158; wildlife in, *80–81, 156–57*
recycling, *146–47*, 147
Red Sea, *39*, 82–84, *82, 83*
reforestation, 158–60, *158, 159*, 161
refrigerators, 98, *99*, 146
Revelle, Roger, 91–92
rice paddies, *64*, 95, *96*
Rocky Mountain Institute, 146, 147, *162*

Sahara Desert, 38–39, *40*
salinity, of oceans, 112
Sally Rock Point (R.I.), *12–13*, 15–16, *28*, 29
San Jose Mercury News, 36
savannas, *25*, 39
Schneider, Stephen, 115
Scott, Senio C., 77
sea level, 18, 49, 110–13, 121–23; coastal encroachment and, 18, *128*, 130–33, *131–33;* islands and, 128–30, *129*
seal, northern, *24*
seasons, 33–36, *34–35*, 105, 108
sedimentary rock, 38–39
Setzer, Alberto, 78–82
smog, *84*
snow, 27, 33, 110, 112, 125
solar energy, 33, 36, 63, 107, *142;* utilization of, *142*, 150–51, *152, 153*. *See also* sunlight
solar ovens, *154*
South Pacific Forum, 130
stromatolites, *43*
Sudan, 125
Suess, Hans, 91
sulfur dioxide, 143
summer heat waves, 123–24, *125*
sun, 44, 45
sunlight, 42, 63, *64–65*, 66, 104, 108; reflected by snow and ice, 110, *111*. *See also* solar energy
"Sunshine" project, *152*
sunspots, 36

taxes: carbon, 153–54, 155; gasoline, 148, 149
temperate climate zones, 36, 49; seasons in, *34–35*
temperature, 123, 135; carbon dioxide and, 69–71, *70*, 73, 91, 94–95; global mean or average, 16, 18, 46, 50, 59, 60, *60*, 94, 99, 110, 121, 123
termites, 95
Thomas, Lewis, 13
Thompson, Starley, 60
thunderstorms, *30–32*, 33
tides, 121. *See also* sea level
tiger, saber-tooth, *47*
tornadoes, 33
Trachodon dinosaur, *47*
trade-wind routes, 36
Trans-Amazon Highway, *78*
TreePeople, *159*
tree planting, 158–60, *158, 159*, 161
tropics, 36, *38*, 154. *See also* rain forests
trucks, 148, 149
tundra, 18, 49, 97
Tuvalu, 130
Tyndall, James, 62–63, *63*, 72

ultraviolet radiation, 44, 45
United Nations, 108, 154
United Parcel Service, 150

Venice (Italy), *132*
Venus, *66*, 67–69
Vernadsky, Vladimir, 87
Vikings, 50
visibility, 77
volcanoes, 41, 46, 95

warmest years on record, 59
waste reduction, 147
water pollution, 60
water vapor, 16, 41, 63, 64, 66, 67, *95*
Watkins, James D., 120
weather, *30–32*, 33; day-to-day changeability of, 33, 36; year-to-year variations in, 91
West Antarctic Ice Sheet, 112
wetlands, 136–39, *137*
wilderness, 133–39, *135–37*
wildlife, *24–25, 128, 137, 138*, 139, *139;* in rain forests, *80–81, 156–57*
wind, 18, 36, 39, 104
windows, 146, *162*
wind power, *140–41*, 150, 151
Winsemius, Pieter, 144
wood, as fuel, 73, 154
World Bank, 153, 154, 155
World Meteorological Organization, 108

Yellowstone National Park, *58*, 59, *59*

2–3 Loren McIntyre; 12–13 and 6, 14 Andrew Revkin; 16 Courtesy of the Rhode Island Historical Society; 19 Loren McIntyre; 20 Sebastíao Salgado, Jr./Magnum Photos, Inc.; 21 top Andrew Holbrooke/Black Star, bottom Mark N. Boulton/Bruce Coleman, Ltd.; 22 Reuters/Bettmann; 24–25 Erwin & Peggy Bauer; 26 *The Physical Geography of the Sea,* by M. F. Maury, LL.D., U.S.N., London: Sampson Low, Son, and Co.; 27 Media Services/NASA; 28 Andrew Revkin; 30–31 and 6 Gunter Zeisler/Bruce Coleman, Ltd.; 34 NOAA Photo Library; 33 Steve Solum/Bruce Coleman, Inc.; 34–35 James P. Jackson/Photo Researchers, Inc.; 37 Jeff Foott Productions; 38–39 and front jacket/Media Services/NASA; 40 Mike Yamashita/Woodfin Camp & Associates, Inc.; 41 Jonathan Blair/Woodfin Camp & Associates, Inc.; 42 Kim Taylor/Bruce Coleman, Inc.; 43 Jan Taylor/Bruce Coleman, Ltd.; 44–45 American Museum of Natural History; 47 top and bottom painting by Charles R. Knight/American Museum of Natural History; 48 Media Services/NASA; 49 Paul Singer; 50 Thomas McGovern; 51 George Holton/Photo Researchers, Inc.; 52 Collection Paul Payot, Propriété du Départment de la Haute-Savoie, Conservatoire d'Art et d'Histoire/Cliche Denis Rigault; 53 top Madeleine Le Roy Ladurie/Bibliotèque Nationale, bottom The Granger Collection; 54 American Museum of Natural History; 56–57 and 6 Larry West Photography; 58–59 Erwin & Peggy Bauer; 60 Paul Singer; 62 Elvin McDonald; 63 The Bettmann Archive; 64–65 Paul Singer; 66–67 Media Services/NASA; 68 Michael Morrison; 69 left Simon Fraser/Science Photo Library/Photo Researchers, Inc., right Debra Meese/Department of the Army; 70–71 Paul Singer; 72 Nicholas Devore/Bruce Coleman, Ltd.; 75 © Chuck O'Rear/Woodfin Camp & Associates, Inc.; 76 Loren McIntyre; 78 Claudia Andujar/Photo Researchers, Inc.; 79 Loren McIntyre; 80 top © Doug Wechsler/Academy of Natural Sciences, bottom Carol Hughes/Bruce Coleman, Ltd.; 81 top left Mrs. Waina Cheng Ward/Bruce Coleman, Ltd., top right Dr. John Mackinnon/Bruce Coleman, Ltd., bottom Luiz Claudio Marigo/Bruce Coleman, Ltd.; 82–83 top Andrew Revkin; 83 bottom left Bill Wood/Bruce Coleman, Ltd., bottom right Carl Roessler/Bruce Coleman, Ltd.; 84 © Jim Mendenhall; 85 and 7 James A. Sugar/© National Geographic Society; 86 David T. Hanson; 87 top left and right The Bettmann Archive; 88 top Dr. Norman Myers/Bruce Coleman, Ltd., bottom Paul Singer; 89 Victor Englebert/Photo Researchers, Inc.; 90 Media Services/NASA; 92 David Moss/Scripps Institution of Oceanography; 94–95 Paul Singer; 96 Porterfield-Chickering/Photo Researchers, Inc.; 97 James A. Sugar/© National Geographic Society; 98 Garry D. McMichael/Photo Researchers, Inc.; 99 © Kent Hanson/Dot Pictures; 100 Robert W. Hernandez/Photo Researchers, Inc.; 102 Jeanne White/Photo Researchers, Inc.; 104 Media Services/NASA; 105 NOAA Photo Library; 106 and 7 Jen and Des Bartlett/Bruce Coleman, Ltd.; 109 top NASA/Science Source/Photo Researchers, Inc., bottom left Larry West Photography, bottom right Larry West/Bruce Coleman, Inc.; 111 Robert W. Hernandez/Photo Researchers, Inc.; 113 NASA/Science Source/Photo Researchers, Inc.; 114 and back jacket Hans Reinhard/Okapia/Photo Researchers, Inc.; 116–117 and 7, 118 © Bruno Barbey/Magnum Photos, Inc.; 121 George Holton/Photo Researchers, Inc.; 122 © Tom Yulsman; 123–24 Paul Singer; 125 Andrew Revkin; 126 Didier/Explorer/Photo Researchers, Inc.; 127 © Mike Yamashita/Woodfin Camp & Associates, Inc.; 128 Greg Ochocki/Photo Researchers, Inc.; 129 Leicagraphie M. Serraillier/Rapho/Photo Researchers, Inc.; 131 Bruce Brander/Photo Researchers, Inc.; 132 left © 1983 Robert John Mihovil, right Ronny Jaques/Photo Researchers, Inc.; 133 Stephen J. Krasemann/Bruce Coleman, Ltd.; 134 © Doug Milner/Woodfin Camp & Associates, Inc.; 135 John Shaw/Bruce Coleman, Ltd.; 136 George Hall/Woodfin Camp & Associates, Inc.; 137 top Jeff Foott Productions, bottom Douglas Faulkner/Photo Researchers, Inc.; 138 Farrell Grehan/Photo Researchers, Inc.; 139 Ron Austing/Photo Researchers, Inc.; 140–41 © Kevin Schafer/Hill Photography; 142 Preston Phillips, Architect, Bridgehampton, NY; 145 Andrew Revkin; 146–47 Will McIntyre/Photo Researchers, Inc.; 149 Geoffrey Gove/Photo Researchers, Inc.; 150 General Motors Corporation; 151 Peter Fisher; 152 Shunji Okura/Photo Researchers, Inc.; 153 and 7 Hank Morgan/Science Source/Photo Researchers, Inc.; 154 Burns-Milwaukee, Inc.; 156 top Sullivan and Rogers/Bruce Coleman, Ltd., bottom Erwin & Peggy Bauer/Bruce Coleman, Ltd.; 157 left C. B. and D. W. Frith/Bruce Coleman, Ltd., right K. Worthe/Bruce Coleman, Ltd.; 158 © Carlos Humberto TDC/Contact Press Images, Inc.; 159 left TreePeople, right, 6, and frontispiece Andrew Revkin; 161 © Michael K. Nichols/Magnum Photos, Inc.; 162 Robert Millman/Rocky Mountain Institute; 163 Mary M. Thatcher/Photo Researchers, Inc.; 164 Jeff Foott Productions

Grateful acknowledgment is given for permission to reprint excerpts from the following books: *Celebrations of Life,* by René Dubos (New York: McGraw-Hill Book Co., 1981), reprinted with permission of the René Dubos Center for Human Environments; *The Firmament of Time,* by Loren Eiseley (New York: Atheneum Publishers, 1975), reprinted with permission of Macmillan Publishing Company; *The Lives of a Cell,* by Lewis Thomas (New York: Bantam Books, Inc., 1975), reprinted with permission of Penguin U.S.A.; *The Star Thrower,* by Loren Eiseley (New York: Harcourt Brace Jovanovich, 1979), reprinted with permission of Random House, Inc.